Structural Dynamic Systems Computational Techniques and Optimization

Computer-Aided Design and Engineering

Gordon and Breach International Series
in Engineering, Technology and Applied Science

Volumes 7–15

Edited by Cornelius T. Leondes

Books on **Structural Dynamic Systems Computational Techniques and Optimization**

Previously published in this series were volumes 1–6 on
Medical Imaging Systems Techniques and Applications

Forthcoming in the *Gordon and Breach International Series in Engineering, Technology and Applied Science*

Mechatronics Systems Techniques and Applications

Biomechanical Systems Techniques and Applications

Computer-Aided and Integrated Manufacturing Systems Techniques and Applications

Expert Systems Techniques and Applications

Structural Dynamic Systems Computational Techniques and Optimization

Computer-Aided Design and Engineering

Edited by

Cornelius T. Leondes

Professor Emeritus
University of California
at Los Angeles

Gordon and Breach Science Publishers

Australia • Canada • China • France • Germany • India •
Japan • Luxembourg • Malaysia • The Netherlands •
Russia • Singapore • Switzerland

Amsteldijk 166
1st Floor
1079 LH Amsterdam
The Netherlands

$518.5 : 621.8$

British Library Cataloguing in Publication Data

Structural dynamic systems computational techniques
  and optimization : Computer-aided design
  and engineering. – (Gordon and Breach
  international series in engineering, technology and applied
  science ; v. 7 – ISSN 1026-0277)
  1. Computer-aided engineering 2. Structural dynamics – Data
  processing
  I. Leondes, Cornelius T.
  624.1'7'0285

ISBN 90-5699-642-8

CONTENTS

SERIES DESCRIPTION AND MOTIVATION

Many aspects of explosively growing technology are difficult or essentially impossible for one author to treat in an adequately comprehensive manner. Spectacular technological growth is made stunningly manifest by any number of examples, but, just to note one here, the Intel 486 IBM-compatible PC was first introduced in late 1989. At that time the price of this PC was in the $10,000 range and it was thought to be much too powerful for widespread use. By early 1992, a little more than two years later, the price had dropped to $1,000 and it was felt that much more power was needed, leading directly to the Pentium IBM-compatible PC. A similar price reduction pattern has already been projected for the Pentium computer, and, in view of the recent history of the 486, it is difficult to suggest that the same "power hungry" pattern will not occur again in a similar time span. The Pentium is presently planned as a 1,000-MHz processor to be called the Flagstaff in the year 2000. The CD-ROM is presently evolving to the DVD (Digital Versatile Disk) with data storage capability of a greater order of magnitude. A DVD-ROM can hold a database of all the phone numbers and addresses in the United States, which would normally require multiple CD-ROMs. And the DVD format has room to grow. In any event, these examples and their clear implications with respect to the many application-oriented issues in diverse fields of engineering, technology and applied science and their continuing advances make it obvious that this series will fill an essential role in numerous ways for individuals and organizations.

Areas of major significance will be defined and world-class co-authors identified as contributors for essential volumes in respective areas. These areas will be determined by criteria including:

1. Will volumes fill important textbook voids in respective areas?

2. In some cases, a "time void" for an important area will clearly suggest the need for a volume. For example, the important area of Expert Systems might have a textbook void of several years that "requires" an important new volume.

3. Are these technology areas that simply cannot sensibly be treated comprehensively by a single author or even several co-authors?

Examples of areas requiring important volumes will be carefully defined and structured and might include, as the case arises, volumes in:

1. Medical imaging

2. Mechatronics

3. Computer network techniques and applications

4. Multimedia techniques and applications

5. CAD/CAE (Computer Aided Design/Computer Aided Engineering)

6. FEA (Finite Element Analysis) techniques and applications

7. Computational techniques in structural dynamic systems

8. Neural networks (as might possibly be suggested by a significant time void in the textbook literature)

9. Expert Systems (again, depending on a possible significant time void).

One of the most important aspects of this series will be that, despite rapid advances in technology, respective volumes will be defined and structured to constitute works of indefinite or "lasting" reference interest.

SERIES PREFACE

The first industrial revolution, with its roots in James Watt's steam engine and its various applications to modes of transportation, manufacturing and other areas, introduced to mankind novel ways of working and living, thus becoming one of the chief determinants of our present way of life.

The second industrial revolution, with its roots in modern computer technology and integrated electronics technology — particularly VLSI (Very Large Scale Integrated) electronics technology, has also resulted in advances of enormous significance in all areas of modern activity, with great economic impact as well.

Some of the areas of modern activity created by this revolution are: medical imaging, structural dynamic systems, mechatronics, biomechanics, computer-aided and integrated manufacturing systems, applications of expert and knowledge-based systems, and so on. Documentation of these areas well exceeds the capabilities of any one or even several individuals, and it is quite evident that single-volume treatments — whose intent would be to provide practitioners with useful reference sources — while useful, would generally be rather limited.

It is the intent of this series to provide comprehensive multi-volume treatments of areas of significant importance, both the above-mentioned and others. In all cases, contributors to these volumes will be individuals who have made notable contributions in their respective fields. Every attempt will be made to make each book self-contained, thus enhancing its usefulness to practitioners in a specific area or related areas. Each multi-volume treatment will constitute a well-integrated but distinctly titled set of volumes. In summary, it is the goal of the respective sets of volumes in this series to provide an essential service to the many individuals on the international scene who are deeply involved in contributing to significant advances in the second industrial revolution.

PREFACE

Structural Dynamic Systems Computational Techniques and Optimization

Computer-Aided Design and Engineering

Because of the availability of powerful computers and highly effective computational techniques, computer aided design and engineering of structural dynamic systems has achieved a high level of unparalleled effectiveness and importance in this area of major significance — so pervasive in applications on the international scene. There are several aspects of the broad process of computer aided design and engineering. First is definition or specification, where criteria for performance of the structural dynamic system are established. Second is synthesis, where a design is created that possibly satisfies criteria established during specification. Third is evaluation, the activity best suited to computer support.

Evaluation can again be divided into three steps. The first step is concerned with representation of the design of a structural dynamic system. This can, for example, be a drawing, mathematical model, geometric model or physical model. The second step involves quantitative analysis of the proposed design — simulation to predict the properties and behavior of the proposed structural dynamic system. The third step compares specifications with results of the analysis. This may necessitate changing the proposed design, but it may also require a change in specifications. During evaluation the proposed design is analyzed to determine how it will function in its working environment. A structural system may, for example, be subjected to stress, heat, vibrations or other relevant environmental factors. If analysis shows unsatisfactory behavior under these conditions, the design must be revised following a procedure similar to that noted above.

This is the second set of volumes in the *Gordon and Breach International Series in Engineering, Technology and Applied Science*, and it consists of 9 distinctly titled and well-integrated volumes on *Structural Dynamic Systems Computational Techniques and Optimization* that can nevertheless be utilized as individual books. In any event, the great breadth of this field certainly suggests the requirement for 9 volumes for an adequately comprehensive treatment.

The set of volumes on structural dynamic systems treats:

1. Computer-Aided Design and Engineering

2. Finite Element Analysis (FEA) Techniques

3. Optimization Techniques

4. Reliability and Damage Tolerance

5. Techniques in Buildings and Bridges

6. Seismic Techniques

7. Parameters

8. Dynamic Analysis and Control Techniques

9. Nonlinear Techniques.

In the first contribution to this volume, Y. M. Xie and G. P. Steven tell us there has been extensive research in the past three decades focused on structural optimization with particular emphasis on frequency restraints. They present an in-depth treatment of the Evolutionary Structural Optimization (ESO) method for frequency optimization. The ESO method is based on the simple concept that by removing inefficiently utilized material, the residual shape evolves toward an optimum. With ESO, shape and topology optimization are achieved simultaneously. A wide range of static and dynamic problems can and have been most effectively treated with the ESO method. Perhaps the greatest advantage of ESO is its simplicity and substantial effectiveness compared with other structural optimization methods.

Optimization of the dynamic behavior of structural dynamic systems brings together some of the more computationally intensive processes in engineering (chapter 2). Optimizing any system by nonlinear programming methods requires repeated and interactive computations to determine an optimum, and applying such optimization methods to dynamic structural problems requires many repetitions of dynamic simulations. Each dynamic simulation in itself may absorb considerable computer time and an overall optimization can require a very substantial and perhaps prohibitive computational effort. Many approaches have been proposed in attempts to reduce required computational time for such optimizations. Some approaches have been successful in reducing computational demands but have, as a result, increased the effort and sophistication required for user preparation. This chapter by A. Narayanan, T. Singh and R. W. Mayne treats some of the issues involved in both efficient optimization of dynamic structural systems and the general movement toward automating this process and making it more effective and routine.

Edwin Boender, Willem F. Bronsvoort and Mark F. Hermeling authored chapter 3. The FE (Finite Element) method computes an approximate solution for a physical problem: given the governing equations for the problem, a computational domain over which the variable must be computed and a sufficient number of boundary conditions. Essential for the FE method is decomposition of the computational domain into a number of finite elements to a sufficient degree of numbers that represent the problem at hand to an adequate degree, in order to achieve necessarily accurate and effective results. Depending on the computational domain dimension, elements can be either 1D, 2D or 3D, and can be embedded within a space of the same or higher dimension. The creation of an FE mesh dividing the computational domain into elements of sufficient number to achieve an adequate accuracy in the final results is a major bottleneck of FE methods. This chapter discusses methods for automatic mesh generation with a view toward achieving design results of adequate accuracy and effectiveness.

A basic and pervasive element in structural dynamic systems is that of thin elastic structures (chapter 4). Generally, explicit solutions of nonlinear partial differential equations describing behavior of these structural elements are not available due, of course, to the mathematical difficulty involved. Usually, practical approaches taken to achieve approximate solutions of adequate accuracy to this problem fall into various broad categories of techniques. L. Azrar, B. Cochelin, N. Damil and M. Potier-Ferry review these techniques and include illustrative examples of highly effective methods of developing solutions for this problem.

In chapter 5 by S. S. E. Lam and D. J. Dawe, we are told that laminated plates are increasingly utilized as structural components in many diverse applications because of their high specific strength and stiffness. These structural elements are characterized by considerable variation and complication in their configuration, and this flexibility results in complexity in predicting or determining structural behavior of laminated plates under loading, as compared with homogeneous isotropic materials, i.e., non-laminated elements. This chapter describes efficient analytical tools used to determine response of laminates and, in particular, their buckling and post-buckling responses. Numerous illustrative examples of a range of applications are included.

The final contribution to this volume by Ramana V. Grandhi and Shyam Sanjeev Kumar relates that forging is a thermomechanical plastic deformation process of universal application where a workpiece is deformed from a relatively simple geometry to a predetermined complex shape by the application of compressive forces. In metal forming processes like forging, it is necessary to control essential process variables like strain, strain rate and temperature in order to obtain desired microstructural and service properties in the final product. This chapter presents a most effective methodology for designing optimal process parameters for isothermal and non-isothermal forging processes with the goal of maintaining system variables within "favorable" processing windows in the strain-strain rate-temperature space. The techniques presented include nonlinear rigid viscoplastic finite element simulation of the metal forming process with computational optimization techniques of mathematical programming. Because of the great universal significance of the forging process, this chapter is a most appropriate conclusion to this volume.

This volume on computer aided design and engineering techniques in structural dynamic systems clearly reveals the effectiveness and great significance of the techniques available and, with further development, the essential role they will play in the future. All the authors are to be highly commended for the splendid contributions that will provide a significant and unique reference for students, research workers, practitioners, computer scientists and others on the international scene for years to come.

1 EVOLUTIONARY STRUCTURAL OPTIMIZATION WITH FINITE ELEMENTS FOR STRUCTURAL DYNAMIC SYSTEMS

Y.M. XIE[1] and G.P. STEVEN[2]

[1] *Department of Civil and Building Engineering, Victoria University of Technology, PO Box 14428, MMC, Melbourne, Australia*
[2] *Department of Aeronautical Engineering, University of Sydney, NSW 2006, Australia*

1.1. INTRODUCTION

The response of a structure to dynamic excitation depends, to a large extent, on the first few natural frequencies of the structure. Excessive vibration occurs when the frequency of the dynamic excitation is close to one of the natural frequencies of the structure. The optimum design of structures with frequency constraints is of great importance, particularly in the aeronautical and automotive industries. In designing most structures, it is often necessary to restrict the fundamental frequency or several of the lower frequencies of the structure to a prescribed range in order to avoid severe vibration. In other circumstances, for example in designing space vehicles, it is desirable to separate closely spaced frequencies so that the control system can work effectively.

Correspondence: Dr. Mike Xie, Department of Civil and Building Engineering, Victoria University of Technology, PO Box 14428, MMC, Melbourne, Australia. Tel: +61 3 9688 4787 Fax: +61 3 9688 4096.

There has been extensive research focused on structural optimization with frequency constraints in the past three decades. A survey by Grandhi[1] covers the recent developments and applications in this area. It is noted that studies of frequency optimization have been mainly restricted to changing the size of beams or the thickness of plates etc. More recently attempts on the simultaneous shape and topology optimization of general structural dynamic systems have achieved some success, most notably the applications of the homogenization method to frequency optimization problems by Díaz and Kikuchi,[2] Ma *et al.*,[3] Tenek and Hagiwara[4] and Ma *et al.*[5]

This chapter presents the Evolutionary Structural Optimization (ESO) method for frequency optimization. The ESO method developed by Xie and Steven[6–8] is based on the simple concept that by removing material that is not efficiently used, the residual shape evolves towards an optimum. With ESO, shape optimization and topology optimization are achieved simultaneously. To date a wide range of static and dynamic problems have been solved using ESO and the results are in excellent agreement with solutions of other techniques such as the homogenization method. The greatest advantage of the ESO method is its simplicity compared with any other structural optimization method.

To use the ESO method for frequency optimization a structure is modelled by a fine mesh of finite elements. At the end of each eigenvalue analysis, part of material is removed from the structure so that the frequency of the resulting structure will be shifted towards a desired direction. A sensitivity number revealing the optimum locations for such material elimination is derived. This sensitivity number can be easily calculated for each element using the information of the eigenvalue solution. In this chapter the ESO method is applied to various frequency optimization problems, which include maximizing or minimizing a chosen frequency, keeping a chosen frequency constant, maximizing the gap of two frequencies as well as considerations of multiple frequency constraints.

1.2. THE SENSITIVITY NUMBER FOR ELEMENT REMOVAL

To identify the optimum locations for structural modifications, sensitivity derivatives are often needed. Usually the sensitivity analysis is very complicated, particularly for dynamic problems. However, we shall show in this section that a sensitivity number which indicates the effect on a natural frequency when an element is removed, can be easily calculated.

First a structure is discretized into a fine mesh of finite elements. The dynamic behaviour of the structure is represented by the following general eigenvalue problem

$$([K] - \omega_i^2 [M])\{u_i\} = \{0\} \tag{1}$$

where $[K]$ is the global stiffness matrix, $[M]$ is the global mass matrix, ω_i is the ith natural frequency and $\{u_i\}$ is the eigenvector corresponding to ω_i. The natural frequency ω_i and the corresponding eigenvector $\{u_i\}$ are related to each other by the Rayleigh quotient

$$\omega_i^2 = \frac{k_i}{m_i} \tag{2}$$

in which the modal stiffness k_i and the modal mass m_i are defined as

$$k_i = \{u_i\}^T [K] \{u_i\} \tag{3a}$$

$$m_i = \{u_i\}^T [M] \{u_i\} \tag{3b}$$

Suppose that we remove one of the elements from the current structure. The change of the frequency due to such an element removal can be derived from equation (2) as

$$\Delta(\omega_i^2) = \frac{\Delta k_i}{m_i} - \frac{k_i \Delta m_i}{m_i^2} = \frac{1}{m_i}(\Delta k_i - \omega_i^2 \Delta m_i) \tag{4}$$

To obtain the value of $\Delta(\omega_i^2)$ from the information of the previous eigenvalue solution, we assume that the eigenvector $\{u_i\}$ is approximately the same before and after the removal of that element. The assumption that the mode shape does not change significantly in between design cycles has been commonly used in frequency optimization.[1] We therefore have

$$\Delta k_i \approx -\{u_i^e\}^T [K^e] \{u_i^e\} \tag{5a}$$

$$\Delta m_i \approx -\{u_i^e\}^T [M^e] \{u_i^e\} \tag{5b}$$

in which $[K^e]$ and $[M^e]$ are the stiffness and mass matrices of the removed element. The element eigenvector $\{u_i^e\}$ contains the entries of $\{u_i\}$ which are related to the removed element. For example if a 4-noded plane stress element is used $[K^e]$ and $[M^e]$ will be 8×8 matrices and $\{u_i^e\}$ will be a 8×1 vector.

Substituting approximations (5a) and (5b) into equation (4) we obtain the approximation of the change of the frequency due to the removal of an arbitrary element, i.e.

$$\Delta(\omega_i^2) \approx \frac{1}{m_i} \{u_i^e\}^T (\omega_i^2 [M^e] - [K^e]) \{u_i^e\} \tag{6}$$

To decide which elements should be removed from the structure so that the frequency will be shifted towards a desired value, we calculate for each element the following *sensitivity number*

$$\alpha_i \approx \frac{1}{m_i}\{u_i^e\}^T(\omega_i^2[M^e]-[K^e])\{u_i^e\} \tag{7}$$

This sensitivity number is an indicator of the frequency change (or to be more precise the change of the square of the frequency) due to the removal of the element.

In modal analysis it is a common practice that the eigenvector $\{u_i\}$ is normalized with respect to the mass matrix $[M]$, which implies that the eigenvector $\{u_i\}$ is scaled to make the modal mass m_i equal to 1. In this case, the sensitivity number α_i can be simplified by omitting m_i in equation (7). Even when the eigenvector $\{u_i\}$ is not normalized, the modal mass m_i which is the same for every element can still be omitted if we are concerned with only one frequency. However, when multiple frequencies are being considered, for example in the case of maximizing the gap of two frequencies, m_i has to be included in the calculation of the sensitivity number unless all the corresponding eigenvectors have been normalized with respect to $[M]$.

It is noted from equation (7) that the sensitivity number α_i can be easily calculated at the element level once the eigenvalue problem (1) is solved. The computational cost involved in calculating the sensitivity number is nominal when compared with the cost of solving the eigenvalue problem.

Assume that the values of α_i are in the range of $[\alpha_{min}, \alpha_{max}]$, where α_{min} and α_{max} are the minimum and maximum values of α_i found over the whole design domain. Typically α_{min} is less than 0 and α_{max} is greater than 0 as shown in Figure 1. There are three regions in Figure 1 which are of importance to the frequency optimization. These regions are marked as A, B and C. For example, to increase the frequency ω_i it would be most effective, according to the definition of the sensitivity number, to remove those elements whose values of α_i fall in region C. Similarly, to reduce ω_i it would be most effective to remove elements whose values of α_i fall in region A. In other circumstances, the designer might be satisfied with the frequency ω_i of the current structure but wishes to reduce the structural weight while keeping that particular frequency constant (or changed as little as possible). This can be achieved by removing elements whose values of α_i are close to zero (i.e. in region B).

Figure 1. The sensitivity number α_i.

From the definition of the Rayleigh quotient in equations (2) and (3), it can be proved that

$$\sum_{elem} \alpha_i = 0 \qquad (8)$$

namely, the summation of the sensitivity numbers over all the elements is equal to zero.

1.3. EVOLUTIONARY PROCEDURES

Generally it is impossible to obtain the optimum design of a structure in one step from the initial design. The optimization process has to be evolutionary, i.e. only a small amount of material, for example 10 elements or one per cent of the total material, should be removed from the structure at each iteration. The solution procedures for various frequency optimization problems are outlined as follow.

1.3.1. Increase a Chosen Frequency ω_i

Step 1: Discretize the structure using a fine mesh of finite elements.

Step 2: Solve the eigenvalue problem (1).

Step 3: Calculate the sensitivity number α_i using equation (7) for each element.

Step 4: Remove those elements with the highest α_i values.

Step 5: Repeat Step 2 to Step 4 until an optimum is reached.

There are several ways to terminate the evolution in Step 5. The simplest way is by prescribing the maximum iteration number (say 50). A similar strategy is to prescribe the maximum amount of material (say 50%) that is allowed to be removed from the structure. The evolution can also be stopped when the frequency under consideration has reached the prescribed target value.

1.3.2. Reduce a Chosen Frequency ω_i

The solution procedure for reducing a chosen frequency is similar to that in sub-section 1.3. The only difference is in Step 4, which should be re-written for the present case as follows:

Step 4: Remove those elements with the lowest α_i values.

1.3.3. Keep a Chosen Frequency ω_i Constant

When material is removed from a structure, the frequencies of the structure will almost inevitably be changed to some extent. What we are trying to do here is to keep the change in a chosen frequency to be as small as possible while the weight of the structure is being reduced. Again the solution procedure is the same as that in sub-section 1.3, except for Step 4 which should be restated as

Step 4: Remove elements whose absolute values of α_i are the smallest.

1.3.4. Increase the Gap between Two Frequencies

By examining the derivation of the sensitivity number in Section 2, it is easy to derive a similar sensitivity number α_{ij} for $(\omega_j^2 - \omega_i^2)$ as

$$\alpha_{ij} \approx \Delta(\omega_j^2 - \omega_i^2) \approx \alpha_j - \alpha_i \qquad \text{where } j > i \qquad (9)$$

which indicates the effect of an element removal on the change in the distance between ω_i^2 and ω_j^2. Once the sensitivity numbers α_i and α_j for individual modes are calculated using equation (7), the new sensitivity number α_{ij} can be immediately obtained from equation (9). The solution procedure is again the same as that in sub-section 1.3, except that Step 4 needs to be restated as

Step 4: Remove those elements with the highest α_{ij} values.

In practice the designer would be only interested in increasing the gap between two neighbouring frequencies, although the present procedure allows him to maximize the distance between two arbitrarily chosen frequencies.

1.3.5. Optimization with Multiple Frequency Constraints

Structural optimization with multiple frequency constraints is similar to the situation of optimization with multiple static load cases. Optimum design of multiple load case structures has been dealt with using the ESO method by Xie and Steven.[7]

The optimization of a structure with multiple frequency constraints is not much more involved than that with a single frequency constraint. The same solution procedure as in sub-section 1.3 can be adopted, provided that a new rejection criterion is used for Step 4.

Figure 2. A short beam under plane stress conditions.

Suppose that we wish to increase the first three frequencies. After solving the eigenvalue problem, the sensitivity numbers α_1, α_2 and α_3 for the first three modes are calculated for each element using equation (7). In order to have the first three frequencies increased, only elements whose values of α_1, α_2 and α_3 are all greater than zero will be removed. If we are most concerned with the first frequency, Step 4 in sub-section 1.3 can be restated as

Step 4: Remove elements whose values of α_1 are the highest among those which have $\alpha_1 > 0$, $\alpha_2 > 0$, and $\alpha_3 > 0$.

1.4. EXAMPLES

The examples presented in this section are all two-dimensional plane stress problems. The element used here is the 4-noded element. However, the ESO method is not restricted to any particular element and it can be applied equally to three-dimensional problems.

1.4.1. A Short Beam

Figure 2 shows a short beam of dimensions 5m × 1m. The beam is clamped on both sides. Young's modulus $E = 200$ GPa, Poisson's $v = 0.3$, thickness $t = 0.1$m and density $\rho = 7000$ kg/m^3 are assumed. The domain is divided equally into 100×20 square elements. Four elements are removed at the end of each finite element analysis.

As the first test we try to increase the fundamental (first) frequency of this short beam. Figure 3 shows the evolutionary history of the first four frequencies of the beam during the first 50 iterations when the first frequency is being increased. This is achieved by removing, at each iteration, four elements whose values of α_1 are the highest among all the remaining elements in the current structure. A new design of the beam corresponding to the 50*th* iteration is given in Figure 4. At this stage 10% of the total

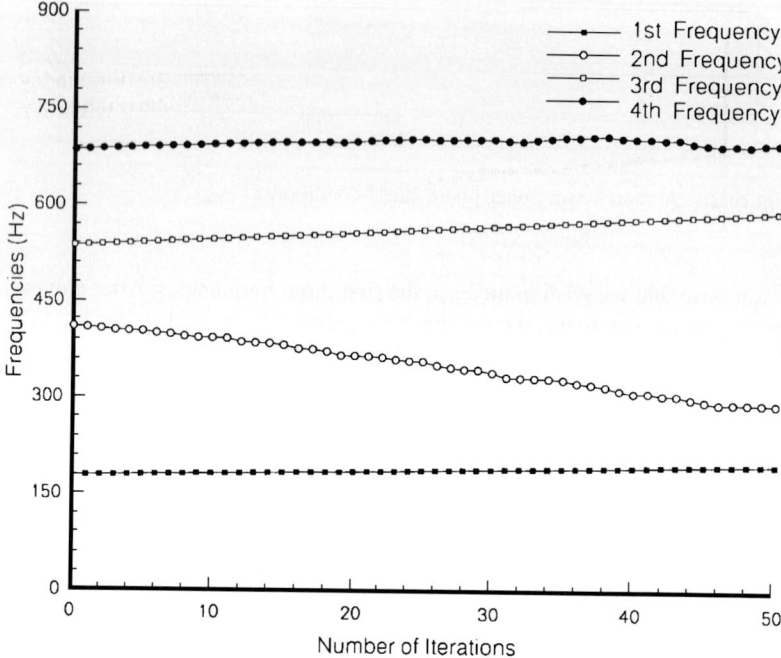

Figure 3. Evolutionary history of the first four frequencies of the short beam: increasing the first frequency.

material has been eliminated and the first frequency has been increased from 178.4 Hz to 199.3 Hz, an increase of 11.7%.

Similarly the second frequency of the beam can be increased by removing elements whose values of α_2 are the highest. The evolutionary history of the first four frequencies for this test is given in Figure 5. After 50 iterations a new design is obtained as shown in Figure 6. The material reduction is again 10% and the second frequency is increased by 9.5% from 409.9 Hz to 448.7 Hz.

Figure 4. A new design of the short beam with increased first frequency. Material removed: 10%; increase in ω_1: 11.7%.

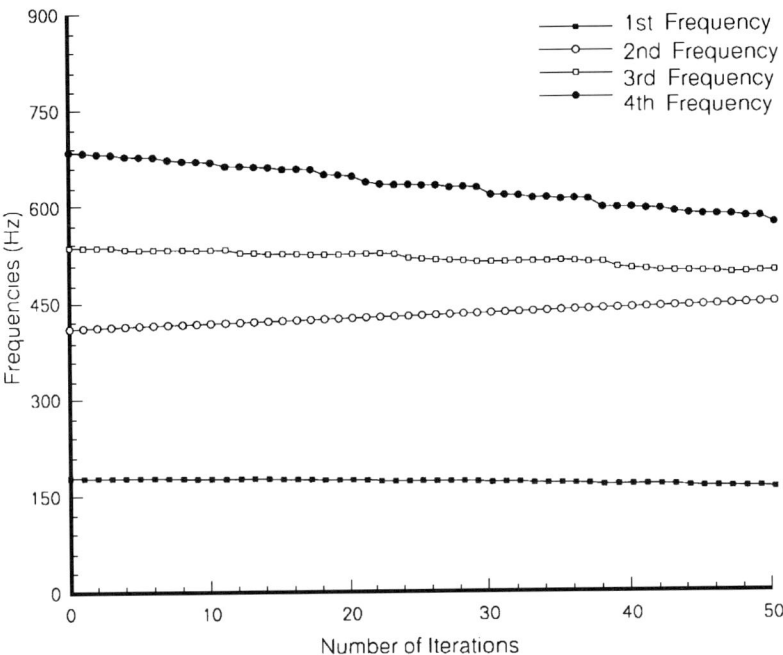

Figure 5. Evolutionary history of the first four frequencies of the short beam: increasing the second frequency.

It is interesting to note that the new shapes and topologies of the short beam shown in Figures 4 and 6 are very similar to what Tenek and Hagiwara[4] obtained using the homogenization method.

Next we consider increasing the gap between the first and the second frequencies of the short beam. Unfortunately no results are available for comparison from other methods in this case, although Ma *et al.*[3] did outline an approach to maximizing the gap between two frequencies using the homogenization method.

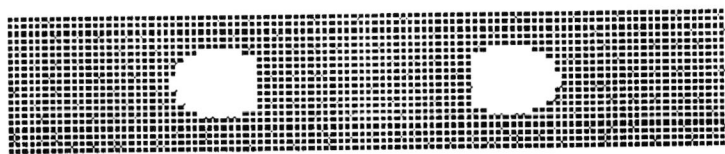

Figure 6. A new design of the short beam with increased second frequency. Material removed: 10%; increase in ω_2: 9.5%.

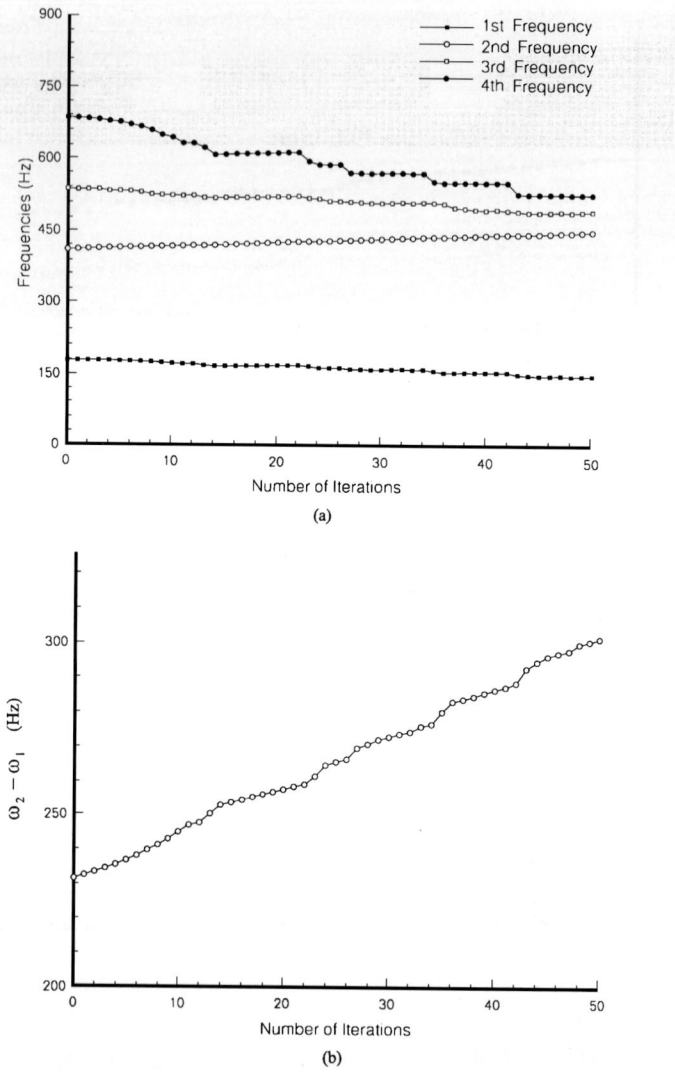

Figure 7. Evolutionary histories for increasing the gap between ω_1 and ω_2 of the short beam. (a) the first four frequencies; (b) the gap between ω_1 and ω_2.

To increase the gap between ω_1 and ω_2, we remove, at each iteration, four elements whose values of $(\alpha_2 - \alpha_1)$ are the highest. Figure 7(a) shows the evolutionary history of the first four frequencies. The increase in $(\omega_2 - \omega_1)$ is shown more clearly in Figure 7(b). After 50 iterations, 10% of the total

Figure 8. A new design of the short beam with increased $(\omega_2 - \omega_1)$. Material removed: 10%; increase in $(\omega_2 - \omega_1)$: 30.1%.

material is removed and $(\omega_2 - \omega_1)$ is increased by 30.1% from 231.5 Hz to 301.2 Hz. The new design is shown in Figure 8.

1.4.2. Rectangular Plate

Figure 9 shows an aluminium plate of dimensions $0.15m \times 0.1m$. The plate is fixed at two corners on its diagonal. Only in-plane vibration is considered here. Young's modulus $E = 70$ GPa, Poisson's $v = 0.3$, thickness $t = 0.01m$ and density $\rho = 2700$ kg/m³ are assumed. The domain is divided equally into 45×30 square elements. Eight elements are removed at the end of each finite element analysis.

Using the procedure outlined in sub-section 1.3 the first frequency can be increased by removing the elements with the highest values of α_1. As a result the evolutionary history of the first three frequencies is obtained as

Figure 9. A rectangular plate under plane stress conditions.

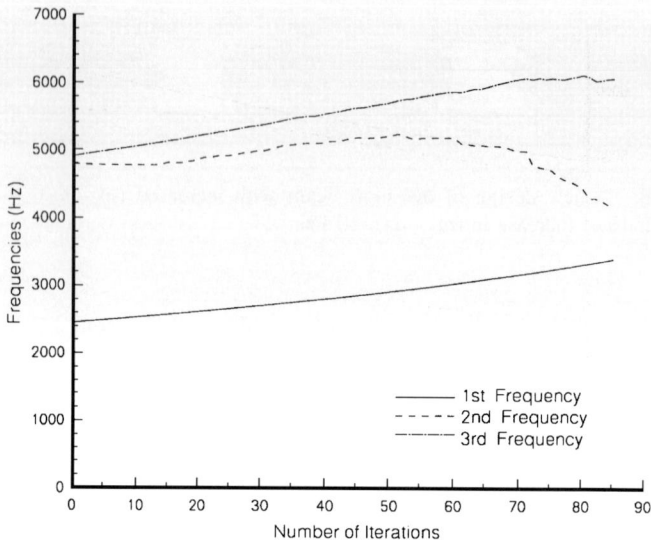

Figure 10. Evolutionary history of the first three frequencies of the rectangular plate: increasing the first frequency.

shown in Figure 10. After 85 iterations, 50% of the total material is removed and the first frequency is increased by 39.0% from 2441.5 Hz to 3392.8 Hz. The corresponding new design of the rectangular plate is given in Figure 11, which is almost identical to what Tenek and Hagiwara[4] obtained using the homogenization method.

In the previous test, only the first frequency has been considered for optimization. There was no control over what is happening to other frequen-

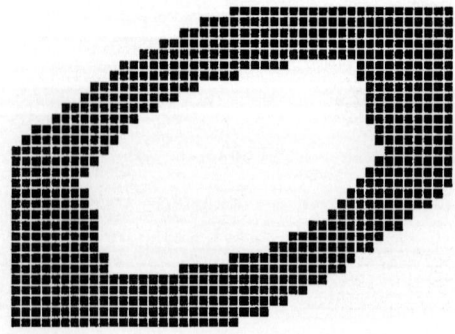

Figure 11. A new design of the rectangular plate with increased first frequency. Material removed: 50%; increase in ω_1: 39.0%.

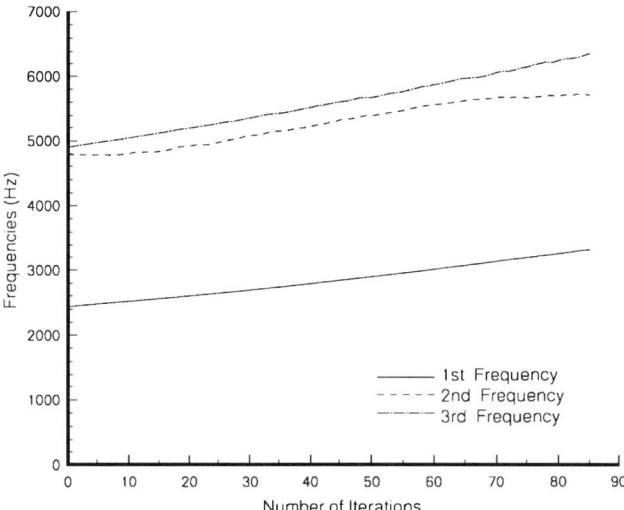

Figure 12. Evolutionary history of the first three frequencies of the rectangular plate: increasing the first three frequencies.

cies during the optimization process. It is clear from Figure 10 that while the first frequency has been increasing monotonically, the second frequency started to drop after about 50 iterations and its value at the 85*th* iteration is even smaller than its initial value. Sometimes, such a result is undesirable.

This can be avoided, or at least delayed, by using the procedure for optimization with multiple frequency constraints as outlined in sub-section 1.3.5. The results for the rectangular plate are shown in Figures 12 and 13.

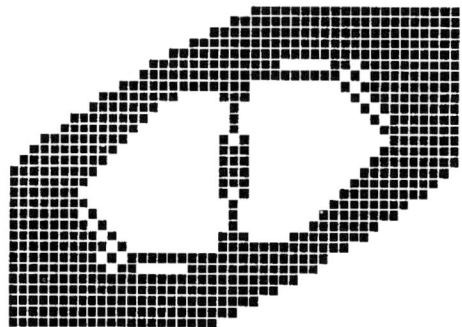

Figure 13. A new design of the rectangular plate with increased first three frequencies. Material removed: 50%; increase in ω_1: 35.5%; increase in ω_2: 19.1%; increase in ω_3: 29.3%.

Figure 14.   A square plate under plane stress conditions.

All the first three frequencies have been increasing for the first 85 iterations, except that ω_2 has slight oscillations towards the end. The new design in Figure 13 has an extra bar in the middle compared with the solution in Figure 11. Again 50% of the total material is removed, the first three frequencies are increased by 35.5%, 19.1% and 29.3%, respectively.

1.4.3. Square Plate

A square plate of dimensions 1m × 1m is clamped on three sides as shown in Figure 14. Young's modulus $E = 200$ GPa, Poisson's $v = 0.3$, thickness $t = 0.01$m and density $\rho = 7000$ kg/m^3 are assumed. The domain is divided equally into 50×50 square elements. Eight elements are removed at the end of each finite element analysis.

First we try to increase the fundamental frequency. Figure 15 shows the new shape of the square plate after 25 iterations when 8% of the total material is removed. The first frequency is increased by 14.3% from 2024.1 Hz to 2313.3 Hz.

Figure 16 is the solution for reducing the first frequency at the $25th$ iteration. The material reduction is again 8%, but the first frequency is reduced considerably from 2024.1 Hz to 162.7 Hz, a decrease of 92.0%. This is due to the narrow "neck" connecting the top and bottom parts, making the bottom part of the structure much more flexible.

Figure 15. A new design of the square plate with increased first frequency. Material removed: 8%; increase in ω_1: 14.3%.

Figure 16. A new design of the square plate with reduced first frequency. Material removed: 8%; reduction in ω_1: 92.0%.

Figure 17. A new design of the square plate with minimum change in the first frequency. Material removed: 8%; change in ω_1: –6.7%.

Our next attempt is to keep the change to the first frequency as little as possible using the procedure outlined in sub-section 1.3.3. Figure 17 shows the result after 25 iterations. The change (reduction) in the first frequency is 6.7% after 8% of the total material has been removed.

1.4.4.   A Short Beam with a Lumped Mass

Similar to the first example, a short beam is clamped at both ends as shown in Figure 18(a). In this example, however, there exists a concentrated mass at the centre of the beam. Details of this example can be found in Zhao *et al.*[9] The optimum design obtained by the ESO method for this short beam is shown in Figure 18(b). It is interesting to note that a similar result has been obtained by Ma *et al.*[5] using the homogenization method.

1.5.   CONCLUDING REMARKS

In this chapter we have demonstrated that the ESO method can effectively solve a wide range of frequency optimization problems. Compared with other existing methods, the greatest advantage of the ESO method is its

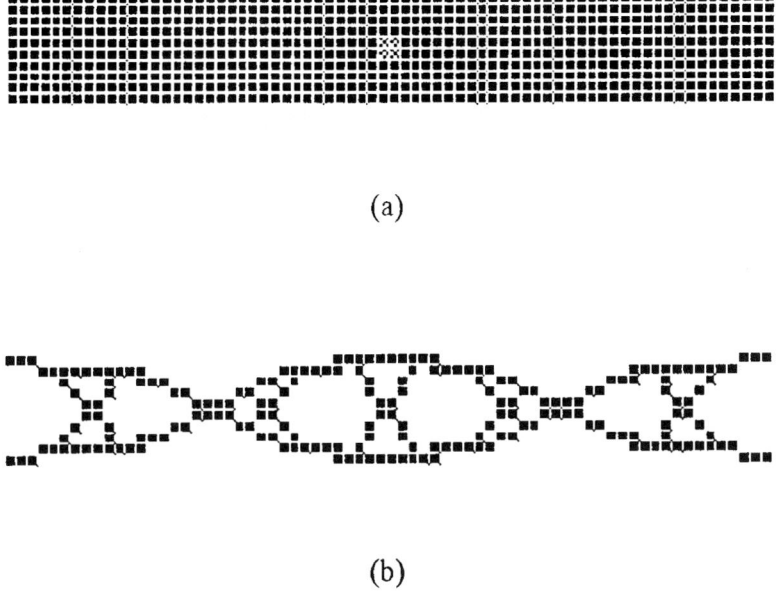

(a)

(b)

Figure 18. Optimization of a short beam with a lumped mass. (a) design domain; (b) optimum design.

simplicity. After three decades of research and development, the Finite Element Analysis has become widely accepted in the engineering community, but unfortunately Structural Optimization does not share the same level of popularity. This situation is caused, at least partly, by the difficulties and complexity suffered by the traditional structural optimization methods. The ESO method has the potential to bridge the gap between the Finite Element Analysis and Structural Optimization. In fact, standard finite element codes can be easily adapted for ESO for frequency optimization by just adding a few lines to calculate and compare the values of α_i from equation (7) for the elements.

Most recently the ESO method has been extended to displacement optimization[10] and buckling load optimization.[11] The simple ESO concept of gradually removing inefficient material proves to work for a whole range of structural optimization problems.

The number of elements to be removed at each step is worth further investigation. In the examples we have considered in this chapter the number of elements removed at each iteration corresponds to 0.2% to 0.6% of the total initial material.

For symmetrical structures, such as those in all the examples of this chapter, one should use an even number for the number of elements removed at each step, in order to keep all the new designs symmetrical. For structures which are symmetrical with respect to two axes, such as the short beam in the last example, the number of elements removed at each step should be a multiple of 4 in order to keep all the new designs double symmetrical. However, one should be aware that if an unsymmetrical finite element mesh is used for a symmetrical structure, the new designs cannot be guaranteed to be symmetrical.

ACKNOWLEDGMENTS

This research is supported by a grant to the authors from the Australian Research Council Large Grants Scheme. This support is gratefully acknowledged. The authors also wish to thank Dr. Chongbin Zhao for his contribution to one of the examples reported in this chapter.

References

1. Grandhi, R.V. 1993, 'Structural optimization with frequency constraints — a review', *AIAA J.*, **31**(12), 2296–2303.
2. Díaz, A.R. and N. Kikuchi, 1992, 'Solutions to shape and topology eigenvalue optimization problems using a homogenization method'. *Int. J. Numer. Methods Eng.*, **35**, 1487–1502.
3. Ma, Z.-D., N. Kikuchi, H.-C. Cheng and I. Hagiwara, 1993, 'Topology and shape optimization methods for structural dynamic problems'. In P. Pedersen, editor, *Optimal Design with Advanced Materials*, pp. 247–261. Elsevier Science Publishers.
4. Tenek, L.H. and I. Hagiwara, 1993, 'Static and vibrational shape and topology optimization using homogenization and mathematical programming'. *Comput. Methods Appl. Mech. Eng.*, **109**,143–154.
5. Ma, Z.-D., N. Kikuchi and H.-C. Cheng, 1995, 'Topological design for vibrating structures'. *Comput. Methods Appl. Mech. Eng.*, **121**, 259–280.
6. Xie, Y.M. and G.P. Steven, 1993, 'A simple evolutionary procedure for structural optimization'. *Comput. Struct.*, **49**(5), 885–896.
7. Xie, Y.M. and G.P. Steven, 1994, 'Optimal design of multiple load case structures using an evolutionary procedure'. *Engineering Computations*, **11**, 295–302.
8. Xie, Y.M. and G.P. Steven, 1994, 'A simple approach to structural frequency optimization'. *Comput. Struct.*, **53**(6), 1487–1491.
9. Zhao, C., G.P. Steven and Y.M. Xie, 1995, 'Evolutionary natural frequency optimization of two dimensional structures with nonstructural lumped masses'. *Engineering Computations*, (submitted).
10. Chu, D.N., Y.M. Xie, A. Hira and G.P. Steven, 1995, 'Evolutionary structural optimization for problems with stiffness constraints'. *Finite Elements in Analysis and Design*, (to appear).
11. Manickarajah, D., Y.M. Xie and G.P. Steven, 1995, 'A simple method for the optimization of columns, frames and plates against buckling'. *Proc. Int. Conf. on Structural Stability and Design*, Sydney, Oct. 30–Nov. 1.

2 ON AUTOMATING THE PARAMETER OPTIMIZATION OF STRUCTURAL SYSTEMS FOR DYNAMIC BEHAVIOR

A. NARAYANAN, T. SINGH and R.W. MAYNE

Department of Mechanical and Aerospace Engineering, University at Buffalo, State University of New York, Buffalo, NY, USA

2.1. INTRODUCTION

Optimization of the dynamic behavior of structural systems brings together some of the most computationally intensive processes in engineering. Optimizing any system by nonlinear programming methods requires repeated and iterative calculations to locate an optimum and applying such optimization methods to dynamic structural problems inherently requires many repetitions of dynamic simulations. Each dynamic simulation in itself may absorb considerable computer time and an overall optimization can require a very substantial and perhaps prohibitive computational effort. Many strategies have been proposed in attempts to reduce the required computational time for such optimizations. Some of these strategies have been successful in reducing computational demands but in return have increased the effort and the sophistication required for user preparation.

The purpose of this chapter is to address some of the issues involved in both the efficient optimization of dynamic structural systems and the general movement toward automating the process to make it more effective and routine. We will briefly discuss the optimization problem and optimization strategies, pointing out important characteristics of the major solution strategies. Different forms of structural system problems will be described and a few overall approaches to optimization will be mentioned which have met a degree of success in reducing the computational effort for these complex problems. One of the most efficient approaches for executing dynamic structural optimization is based on the application of sensitivity analysis to the dynamic system equations in order to establish derivative information for either eigenfrequency values or for measures of transient system behavior. This derivative information is critical to the operation of the most efficient gradient based optimization methods. However, obtaining accurate derivative information by sensitivity analysis has the disadvantage of generally requiring increased skill and increased preparation time on the part of the user.

The focus of this chapter is on the use of symbolic manipulation to provide a computer based tool for automating the sensitivity analysis process and integrating this procedure with optimization strategies and system analysis techniques. This approach has the advantage of employing efficient gradient based optimization algorithms without requiring the user to manually prepare algebraic gradient information. The initial efforts described employ Macsyma for symbolic manipulation. The concluding discussion considers Maple and shows the use of symbolic manipulation in developing finite element equations well suited for sensitivity analysis and optimization.

2.2. NONLINEAR PROGRAMMING

The traditional nonlinear programming problem is to

$$\text{minimize} \quad f(z) \tag{1}$$
$$\text{wrt } z$$

$$\text{with} \qquad g_i(z) \leq 0 \quad i = 1, \ldots, p \tag{2}$$

$$\text{while} \qquad h_j(z) = 0 \quad j = 1, \ldots, q \tag{3}$$

where z contains the design variables to be adjusted during the optimization, $f(z)$ is the objective function, the $g_i(z)$ are the inequality constraints and the $h_j(z)$ are the equality constraints. This problem is defined in a most general form and has the potential to have objective and constraint functions which are very poorly behaved, highly nonlinear, discontinuous, etc.

Of course, a wide variety of optimization techniques have been proposed to solve the optimization problem. These methods range from intuitively based search strategies such as the "pattern search" (Hooke and Jeeves, 1961) and "flexible polyhedron" (Nelder and Mead, 1964; also Himmelblau, 1972: 148–157) approaches to techniques with a firm analytical basis such as the quasi-Newton methods for unconstrained optimization (Reklaitis *et al.*, 1983: 112–116), multiplier penalty functions (Schuldt, 1975), generalized reduced gradient (GRG) methods (Abadie and Carpentier, 1969) and sequential quadratic programming (SQP) algorithms (Powell, 1978). A number of comparative studies were performed and continue to be performed to evaluate the effectiveness of various optimization techniques. The studies by Sandgren and Ragsdell (1980) and Schittkowski (1982) were especially detailed. Following their publication it became widely accepted that the analytically based approachesof GRG and particularly SQP had very substantial advantages for solving many realistic optimization problems. These advantages appeared in terms of robustness and solution accuracy but especially in terms of a reduced number of objective function and constraint evaluations required to complete an optimization. The number of such function evaluations is normally viewed as a direct measure of the amount of computational effort used in the optimization.

In more recent years, the treatment of problems with very poorly behaved objective functions and complex constraints has received increasing attention. A variety of optimization methods in the forms, for example, of simulated annealing (Corana *et al.*, 1987) and genetic algorithms (Goldberg 1989) have been developed with an ability to find the global optimum in an almost pathological environment including discontinuous and multimodal functional relationships. The penalty for this tenacious ability for locating global optima is that a large number of function evaluations is typically necessary. Applying these global methods to more reasonably conditioned problems can result in excessive numbers of function evaluations compared to GRG or SQP algorithms. For global optimization with an engineering premise, some comparative information is available in Borup and Parkinson (1992) where four different strategies are compared in the optimization of poorly behaved but realistic engineering problems. In that comparison, a variation of one of the early intuitively based optimization methods (Nelder and Mead, 1964) turned in very respectable performance compared to the simulated annealing and genetic approaches. For optimization problems which are reasonably well conditioned, however, there is still no reason to doubt the conclusion of Schittkowski (1982) that GRG and SQP are superior especially in terms of computational efficiency. Both of these approaches require extensive gradient information for their use.

2.3. FORMULATIONS FOR DYNAMIC SYSTEM OPTIMIZATION

Dynamic system optimization problems can be posed in a state equation format. For linear systems

$$\underline{\dot{x}} = A(\underline{z})\underline{x} + B(\underline{z})\underline{u} \qquad (4)$$

describes the dynamic character of the system where the behavior of the state variables x depends on the variables in the design vector z which define the coefficient matrices A and B. These design variables may represent the stiffness or damping coefficients in a mechanical system, the gains of a closed-loop control system or, in the case of a structural design problem, the geometry and dimensions of structural members. The objective function and constraints may include direct algebraic functions of the design variables in z such as the weight of a structure but may also include a variety of dynamic functions such as a minimum allowable value for the system eigenvalues, an integral time absolute error measure of transient response, a maximum allowable stress value during a transient, etc. The possibilities are endless and, of course, a second order formulation of the system dynamics may also be useful instead of the first order state equations shown above. In cases where nonlinear effects are important in the dynamic behavior of the system, these may be included in the more general state equation form:

$$\dot{x} = f(\underline{x}, \underline{z}, \underline{u}, t) \tag{5}$$

Let's consider that a representative dynamic design optimization problem may be to adjust the parameters of a structural system to obtain the fastest possible transient response while meeting a variety of constraints such as a maximum acceptable weight, a maximum static deflection and a maximum acceleration level during the transient. In a straight-forward approach to this problem, the state equations for the system can be integrated numerically to evaluate the dynamic behavior and with a few "IF" statements, the maximum values during the transient could be monitored. In this way, the system performance can be determined for a given set of design parameters. With repeated simulations and adjustments of the design parameters, an optimization algorithm will make progress toward the optimum. By making perturbations in z, finite difference gradient information can be obtained and more effective algorithms such as SQP can be employed for the optimization.

This straight-forward approach to simulation/optimization has the advantage of requiring relatively little user preparation beyond formulating the simulation itself and coupling it to an optimization algorithm. The disadvantage of the strategy is that the computer time required with most optimization algorithms will be considerable and perhaps prohibitive for many real problems. Choice of an SQP algorithm may be helpful to reduce computational effort but finite difference derivatives have only limited accuracy and could interfere with the convergence of the strategy. The repeated function evaluations necessary to obtain finite difference derivatives can also make a very considerable increase in the computational time per iteration of the algorithm.

Similar limitations also occur in attempting to apply frequency domain criteria directly to the dynamic system. The direct calculation of eigenvalues and natural frequencies can be used for parameter optimization. But, there are difficulties with computational effort as problems become large and finite difference differentiation leads to numerical difficulties and/or contributes to the substantial computational effort.

2.4. ALTERNATIVE OPTIMIZATION STRATEGIES

Two basic strategies have emerged in attempts to deal with the excessive computational load posed by dynamic structural optimization problems. One of these strategies has been to avoid using the actual dynamic calculations during the optimization and instead to use a surrogate function which can be evaluated algebraically. The surrogate function is obtained by a variety of fitting approaches based on an initial series of dynamic calculations over a range of design variable values. The other strategy for improving computational times has been to obtain gradient information for use in the optimization by exact derivative calculations. These derivatives are obtained by methods of sensitivity analysis which can produce accurate derivatives and reduce computation time by eliminating the function evaluations needed for finite differences. Of course, the derivative information from sensitivity analysis can be well used in the gradient based optimization procedures known for their efficiency.

Examples of the fitting strategy and the sensitivity analysis strategy are described briefly below. The work which comprises the remainder of this chapter focuses on the sensitivity analysis approach and procedures for implementing it in a user friendly way, using symbolic manipulation to reduce user effort and tedious preparation.

2.4.1. "Fitting" in Structural Problems

Roozen-Kroon *et al.* (1989) considered the shape optimization of bells where the critical issue was the ratio of eigenfrequencies required to yield desired tones. Design variables (as many as twenty) were used to describe the bell wall shape. Because of the complexity of the problem and computations, a direct optimization was considered to be impractical in the context of repeated eigenvalue calculations. To solve this problem, a factorial design of experiment approach was used to obtain data for a regression fit of the eigenfrequencies and eigenfrequency derivatives as functions of the design variables. The dynamic equations were solved for the eigenvalue information at selected design variable values. Based on the resulting regression equations, optimizations were then possible using an SQP algorithm.

The problem of designing structures to reduce radiated noise was considered by Hambric (1995). An objective function coupling radiated noise, weight and manufacturing cost was considered. The expense of the full system solutions necessary for the radiated noise calculations made optimization difficult, particularly since it was desired to apply global optimization solutions. Global optimization typically requires extensive iterative calculation. Various fitting strategies were considered including a quasi-Newton approach for fitting a performance surface to a series of full system solutions. Once the fit of the combined objective was obtained, the simulated annealing global optimization algorithm (Corana *et al.*, 1987) was used to find the optimal structure considering the combined objective and using only the fitted surface to represent the function during optimizations.

Fitting approaches may appear attractive for complex optimization calculations but it is not clear that overall computer time is reduced. The careful study by Rogers and LaMarsh (1995) considered fitting by various neural network approaches. Comparing the computation time results obtained by a direct optimization to the time required for obtaining the fitted data showed only two of five attempted procedures able to obtain a fit faster than the gradient based optimization could be performed. It was found that a fitting procedure could only assure a time saving if parallel computation techniques were used during the fitting process.

2.4.2. Solutions by Sensitivity Analysis

The work presented by Grandhi and Venkayya (1988) illustrates the use of sensitivity analysis in designing structures for dynamic character based on frequency domain behavior. The design of member cross-sections for several different truss structures was presented for up to 200 members with an objective of minimizing structural weight while applying inequality and equality constraints to the natural frequencies. The derivatives of the natural frequencies with respect to the design variables were obtained by a sensitivity approach which avoided finite differences. The optimization strategy was based on a direct minimization of the Lagrangian.

The work of Haug and Arora (1979) applied sensitivity analysis to a variety of transient response problems including mechanical systems, control systems and structural behavior. In the process they created many of the strategies useful for sensitivity analysis in the context of transient behavior. One of the Haug and Arora examples considered a five-degree-of-freedom vehicle suspension model with an objective function of maximum passenger acceleration during a bump and with the maximum wheel displacements limited by constraints. Spring stiffnesses and damper values were to be selected in the optimization. Another example discussed in Haug *et al.* (1986) deals with a vibrating beam and the optimization of the beam cross-section. The objective function is the time integral of the beam displacement squared.

Ashrafiuon and Mani (1990) applied sensitivity analysis to a variable stroke compressor, a system involving both mechanical dynamics and structural concerns. Optimal design methods were demonstrated and applied to smoothing the vibrations of the rotating compressor. Symbolic computations were used to reduce the computational time and to make the generation of sensitivity information practical in the compressor's complex kinematic environment.

2.5. DEVELOPING EQUATIONS FOR SENSITIVITY ANALYSIS

The function fitting strategy and the use of sensitivity analysis are the two methods most frequently considered for the optimization of dynamic structural systems when computer time is an important consideration. These are thought of as the most tractable ways to minimize the inherently large computational effort in such problems. The use of function fitting may be especially attractive for situations where an overview of the entire design space may be required and where global optimization seems to be a necessity. However, in many real problems, sensitivity analysis and the direct optimization of the system behavior can be most appropriate. Function and constraint evaluation can be restricted to the general path leading to optimum. Performing function evaluations which survey the entire design space (as needed in fitting) is not required. For problems where there is the possibility of alternative local minima, starting a local optimization process from a few different starting points will typically locate the global minimum. In fact, successive local searches have been proposed as a strategy very much competitive with actual global optimization (Jain and Agogino, 1993).

The viewpoint taken in this chapter is that gradient based search strategies are an important tool for the optimization of dynamic systems including structures. The efficiency and effectiveness of such searches are enhanced by the accurate derivative information resulting from sensitivity analysis and by avoiding finite differences. The user effort required to develop sensitivity equations for dynamic systems is very significant and may well discourage the more general use of this approach. For this reason the work described here has focused on strategies for automating sensitivity analysis through symbolic manipulation.

2.5.1. Sensitivity Equations for Transient Behavior

In considering the transient performance of dynamic systems, an objective function of the form:

$$J = \int_0^{t_f} \Phi(\underline{x}(t),\underline{z},\underline{u}(t),t)dt + \theta(\underline{z}) \tag{6}$$

permits a wide variety of integral performance measures and also includes a term for static system performance based directly on the design parameters. There also a number of useful constraint forms which are presented below using the format suggested by Narayanan (1989).

Type "A" constraints may be active at a particular instant of time:

$$g_s(\underline{x},\underline{z})\Big|_{t_j} \leq 0 \quad \text{for a specified } t_j \tag{7}$$

Type "B" constraints apply an integral constraint form to capture dynamic system performance:

$$\int_0^{t_f} g_d(\underline{x},\underline{z},\underline{u},t)\, dt \leq 0 \tag{8}$$

Type "C" constraints are useful for monitoring variables critical to system performance during transients and can assure that they do not exceed particular values:

$$g_{ds}(\underline{x},\, \underline{z},\, \underline{u},) \leq 0 \quad \text{for } 0 \leq t \leq t_f \tag{9}$$

The forms above are shown as inequality constraints, they may also take on equality forms as necessary. The type A constraints are viewed as "static" (subscript s) since they apply to a specific instant of time. The type B constraints are "dynamic" (subscript d) and are integrated over an interval of time. The type C constraints (subscript ds) are considered to be "quasi-dynamic" and are triggered by exceeding their allowable value during the transient. Notice that the objective function of Equation 6 combines the static and dynamic forms of types A and B. Overall, the three constraint representations permit many situations to be considered. The type A constraint can allow a limitation on structural weight or a static deflection, for example. The type B constraint can permit limitations on control effort over a desired time interval and the type C constraint will permit maximum limits to be placed on acceleration or velocity during a transient.

Obtaining constraint and objective function derivatives in the transient environment can be performed by a direct differentiation approach. The dynamic constraints are the most difficult to consider and are the source of complications in transient response optimizations. For a dynamic constraint of type B, consider

$$G = \int_0^{t_f} g_d(\underline{x}, \underline{z}, \underline{u}, t)\, dt \qquad (10)$$

To obtain the derivative with respect to the design variables, we differentiate Equation 10, yielding

$$\frac{dG}{d\underline{z}} = \int_0^{t_f} \frac{dg_d}{d\underline{z}}(\underline{x}, \underline{z}, \underline{u}, t)\, dt \qquad (11)$$

If we assume that the input does not play a part in the design process

$$\frac{dg_d}{d\underline{z}} = \frac{d\underline{x}^T}{d\underline{z}} \frac{\partial g_d}{\partial \underline{x}} + \frac{\partial g_d}{\partial \underline{z}} \qquad (12)$$

The derivative of the state variables with respect to the design variables can be obtained by differentiating the system state equations and appropriately integrating. For example, in the case of linear systems, differentiation yields

$$\frac{\partial \underline{\dot{x}}}{\partial z_i} = A \frac{\partial \underline{x}}{\partial z_i} + \frac{\partial A}{\partial z_i} \underline{x} + \frac{\partial B}{\partial z_i} \underline{u} \qquad (13)$$

for the ith design variable. The equations above can be integrated together to obtain the desired gradient vector $dG/d\underline{z}$ over the transient response period. The role of symbolic manipulation in this process is to develop Equations 12 and 13 algebraically, based on the descriptions of $A(\underline{z})$, $B(\underline{z})$ and g_d. Nonlinear systems can be treated similarly using the state equation form of Equation 5. The strategy also extends readily to the treatment of control inputs to the system $\underline{u}$ as part of the design process. This allows input optimization problems to be structured and feedback control arrangements to be considered. Equations 10–13 apply to the type B constraints described above. Type A constraints can be handled more directly, requiring differentiation but not requiring integration. Type C constraints can be treated conveniently by using the representation

$$G = \int_0^{t_f} < g_{ds}(\underline{x}, \underline{z}, \underline{u}, t) > dt \qquad (14)$$

where

$$< g_{ds} > \; = 0 \qquad \text{for } g_{ds} \leq 0$$
$$= g_{ds} \qquad \text{for } g_{ds} > 0$$

This then allows the integral formulation of the type B constraints to be extended to constraints which limit the maximum value of variables or functions during the transient.

2.5.2. Sensitivity Equations for Eigenvalues

In certain design problems, the natural frequencies of the system are explicitly important and are of interest in the design process. In other cases, system eigenvalues are a convenient way of characterizing the transient behavior of a system and may be used in optimal design procedures to avoid the overhead associated with performing actual transient response computations during the optimization. Gradient based search methods may be helpful in eigenvalue optimization strategies and can be facilitated by direct calculation of eigenvalue sensitivities.

A convenient procedure for obtaining eigenvalue sensitivities has been developed in Junkins and Kim (1993) and is based on the "nondefective" state equation representation

$$B \, \dot{\underline{x}} = A \, \underline{x} \tag{15}$$

This form is particularly attractive for structural dynamics problems and in the conversion from a second order dynamic system formulation. The eigenvalue sensitivity to a system design variable is shown to be

$$\frac{\partial \lambda_j}{\partial z_i} = \underline{\psi}_j^T \left(\frac{\partial A}{\partial z_i} - \lambda_j \frac{\partial B}{\partial z_i} \right) \underline{\phi}_j \tag{16}$$

where

$$\lambda_j = \text{the } j\text{th eigenvalue}$$
$$z_i = \text{the } i\text{th design parameter}$$
$$\underline{\psi}_j = \text{the } j\text{th left eigenvector}$$
$$\underline{\phi}_j = \text{the } j\text{th eigenvector}$$

Equation 16 can be used to calculate eigenvalue sensitivities based on eigenvalue and eigenvector information and the sensitivities of the A and B matrices to the design variables. Obtaining the matrix sensitivities during computations can be expedited by using symbolic manipulation to develop algebraic representations for the derivatives. For given values of the design variables, a direct substitution can then be used for the sensitivity computation.

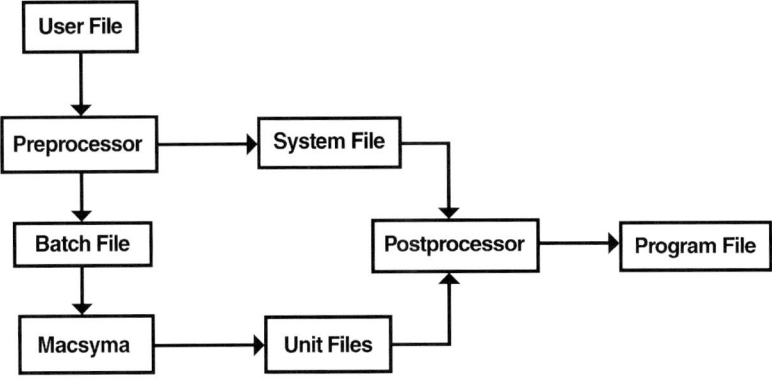

Figure 1. An automation of sensitivity analysis.

2.6. AUTOMATION OF THE TRANSIENT DESIGN PROCESS

The major difficulty in using sensitivity analysis techniques in the design optimization process is that the derivatives of large numbers of complicated expressions must be obtained. This differentiation may make it possible to obtain accurate sensitivities and to permit robust and computationally efficient design optimization. However, it increases user effort and prevents the typical engineering designer from realistically using sensitivity analysis methods in the optimization process. Our initial efforts to automate the design of dynamic systems for transient behavior (Narayanan, 1989; Narayanan and Mayne, 1989) were directed toward a convenient approach for integrating sensitivity analysis into the optimal design process by using symbolic manipulation. This initial work was restricted to linear systems in the standard state equation form of Equation 4. However, nonlinear objective functions and constraints were permitted.

The strategy shown in Figure 1 is based on the use of the symbolic manipulation package Macsyma (Symbolics Inc., 1985). In this strategy, the user prepares a "User" file which describes the system of interest in terms of its dynamic state equations as well as the objective function and constraints which define the design problem. A Fortran Preprocessor then reads the User file as character strings. It processes this input and prepares two output files. One of these files (the "Batch" file) is a file of Macsyma commands defined by alphanumeric output of the Preprocessor. This file can be read by Macsyma which can then apply symbolic differentiation as necessary to create exact derivative information for the optimization. The other file created by the Preprocessor is a "System" file which is used at the

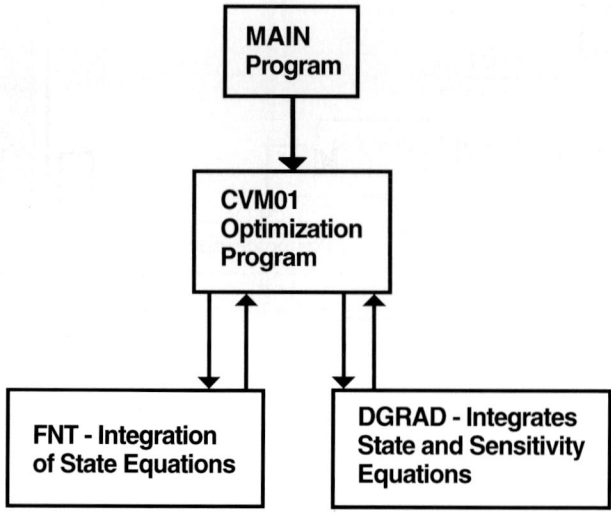

Figure 2.   Structure of the optimization process.

next stage of the formulation strategy. This file contains specific numerical information needed for the system optimization including the length of the simulation interval, the initial conditions for the dynamic system solution and the dimensions of the problem.

After running the Preprocessor, Macsyma is executed using the Batch file. The Macsyma execution then creates the "Unit" files which form the basis for the function evaluation and derivative evaluation subroutines required by the optimization programs. The Unit files contain the algebraic information necessary to define the objective function and its gradient, the individual constraints and their gradients, the state equations and the sensitivity equations for $\partial \underline{\dot{x}} / \partial z_i$. The third step in this formulation process is the execution of a "Postprocessor" Fortran program. This program reads the System file prepared by the Preprocessor and reads the Unit files output by Macsyma. It then assembles the program and subroutines ready for use with an optimization package to perform dynamic system optimization. The overall process is similar to the compiling and linking of a few files as necessary to write and execute a computer program. Of course, the procedure could be made quite transparent to the user by executing the successive programs with system level commands.

The overall "Program" file has the structure shown in Figure 2 and is typical of the arrangements used in packages of optimization routines. The "Main" program is created by the Postprocessor based on System file

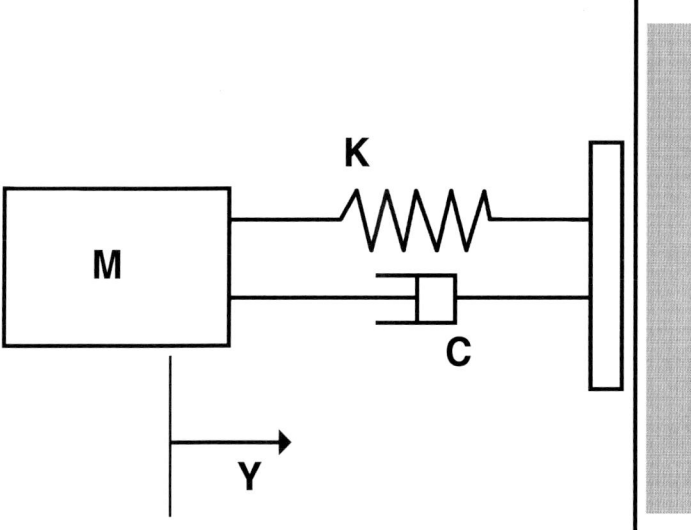

Figure 3.   A simple shock absorber.

information. It calls the optimization subroutine CVM01 which is a sequential quadratic programming algorithm (Wang, 1985). The "Fnt" routine is
used to evaluate the objective function and constraint values for a given set
of design variables. The "Dgrad" routine evaluates the gradients of the
objective and constraints. Both of these subroutines are automatically prepared by the Postprocessor using Batch file information and segments of
Fortran code from the Unit files produced by Macsyma. Integrations included in Fnt and Dgrad are performed by a simple Runge-Kutta approach.

2.6.1. A Simple Example

The demonstration problem for this initial study considered a very basic
shock absorber optimization problem to illustrate the overall procedure for
automatically performing the symbolic manipulation of sensitivity analysis
and generating the Fortran coded program for a transient optimization. This
example was originally considered in Afimiwala and Mayne (1974) and
consists of a mass, spring and damper set up in the bumper configuration
of Figure 3. The system differential equation is, of course,

$$\ddot{y} + C\dot{y} + Ky = 0$$

For the shock absorber transient starting with an initial velocity and zero displacement, the initial conditions are $\dot{y}(0) = 1$, $y(0) = 0$ considering normalized velocity and displacement variables. The optimization of the transient is to adjust K and C to yield minimum peak deceleration with a maximum stroke of unity during the transient. In the state equation format of the automated procedure:

$$\underline{x} = \begin{bmatrix} y \\ \dot{y} \end{bmatrix}$$

with initial conditions

$$\underline{x}(0) = \begin{bmatrix} 0 \\ 1 \end{bmatrix}$$

and

$$A = \begin{bmatrix} 0 & 1 \\ -z_1 & -z_2 \end{bmatrix}$$

Also, since an "input" is expected in the standard form of the transient optimization, let $B = [0 \ 1]^T$ and $u = 0$. The optimization problem is formulated by considering a third design variable to facilitate the handling of the maximum value requirement on deceleration so that the problem statement becomes

$$\text{minimize } z_3$$
$$\textit{wrt } \underline{z}$$

subject to   (17)

$$g_1 = \max x_1 - 1 \leq 0$$
$$g_2 = \max |\dot{x}_2| - z_3 \leq 0$$

The new design variable becomes the objective function and, when minimized, forces the deceleration to its lowest value.

The objective function above is simply in a static form. The constraints are both quasi-dynamic and take on the form of Equation 14 in the optimization process. For example, the first constraint is treated as

$$\int_0^{t_f} < \max x_1 - 1 > dt \leq 0 \qquad (18)$$

```
3,2,1,301,0,0,0,2,0,0,0,2,0
0.,10.
0
1.0
-1.0*z[1]/1.0
-z[2]/1.
0
0
0
1.0e-10
1.
z[3]
(abs(x[1,1])-1.0)
(abs(outvec[2,1])-z[3])
```

Figure 4. User file for the shock absorber optimization.

so that the sensitivity analysis approach of Equations 10–13 applies. The User file prepared to solve the simple shock absorber optimization is shown in Figure 4. It contains only the most basic information. For example, the first line contains the number of design variables, the number of state variables, the numbers of the various kinds of constraints, etc. Lines 3 to 6 show the A matrix and the last three lines define the objective function and the constraints. Figures 5 and 6 show some examples of the Unit file output from Macsyma. Figure 5 shows the sensitivity equations where $xz(i,j) = \partial x_i/\partial z_j$ and $xnext(i,j) = \partial \dot{x}_i/\partial z_j$. The output file of Figure 6, shows the derivative results for the two quasidynamic constraints. It reflects the computation of Equation 12 and presents the components of dg_{ds}/dz as $gdsz(i,j)$.

Of course, the quality of the derivative information in an optimization problem is a continuing concern. In order to gain confidence in the auto-

```
fortran('xnext=float(xnext))$
xnext(1,1) = xz(2,1)
xnext(1,2) = xz(2,2)
xnext(1,3) = xz(2,3)
xnext(2,1) = -1.0*z(2)*xz(2,1)-1.0*z(1)*xz(1,1)-1.0*x(1,1)
xnext(2,2) = -1.0*z(2)*xz(2,2)-1.0*x(2,1)-1.0*z(1)*xz(1,2)
xnext(2,3) = -1.0*z(2)*xz(2,3)-1.0*z(1)*xz(1,3)
```

Figure 5. Basic sensitivity in the unit file.

```
fortran('gdsz=float(gdsz))$
gdsz(1,1) = x(1,1)*xz(1,1)/abs(x(1,1))
gdsz(1,2) = x(1,1)*xz(1,2)/abs(x(1,1))
gdsz(1,3) = x(1,1)*xz(1,3)/abs(x(1,1))
gdsz(2,1) = z(2)*(z(2)*x(2,1)+z(1)*x(1,1))*xz(2,1)/abs(z(2)*x(2,1)
1     +z(1)*x(1,1))+z(1)*xz(1,1)*(z(2)*x(2,1)+z(1)*x(1,1))/abs(z(2)*x
2     (2,1)+z(1)*x(1,1))+x(1,1)*(z(2)*x(2,1)+z(1)*x(1,1))/abs(z(2)*x(
3     2,1)+z(1)*x(1,1))
gdsz(2,2) = z(2)*(z(2)*x(2,1)+z(1)*x(1,1))*xz(2,2)/abs(z(2)*x(2,1)
1     +z(1)*x(1,1))+x(2,1)*(z(2)*x(2,1)+z(1)*x(1,1))/abs(z(2)*x(2,1)+
2     z(1)*x(1,1))+z(1)*xz(1,2)*(z(2)*x(2,1)+z(1)*x(1,1))/abs(z(2)*x(
3     2,1)+z(1)*x(1,1))
gdsz(2,3) = z(2)*(z(2)*x(2,1)+z(1)*x(1,1))*xz(2,3)/abs(z(2)*x(2,1)
1     +z(1)*x(1,1))+z(1)*xz(1,3)*(z(2)*x(2,1)+z(1)*x(1,1))/abs(z(2)*x
2     (2,1)+z(1)*x(1,1))-1.0

closefile()$
```

Figure 6.   Unit file expressions for constraint derivatives.

mation process, comparisons were made between the results obtained by the automated sensitivity analysis calculations and similar results obtained by finite differences. Figures 7 and 8 show typical results for the constraint functions of the shock absorber problem. The calculations by finite differences at various step sizes lack precision (as anticipated) and the sensitivity analysis results have appropriate values.

The optimum shock absorber transient shown in Figure 9 results from solution of the optimization problem prepared by the Postprocessor of Figure 1 using the System file and Unit files produced by the Preprocessor and Macsyma, respectively. The full unit stroke of the shock absorber is utilized with a maximum deceleration level of 0.520. The optimum values

	$\left(\dfrac{dG1}{dZ} \right)$		
	$\dfrac{dG1}{dZ_1}$	$\dfrac{dG1}{dZ_2}$	$\dfrac{dG1}{dZ_3}$
Sensitivity	−6.610	−4.044	0
fd = 0.1	−4.057	−3.327	0
fd = 0.01	−6.214	−3.959	0
fd = 0.001	−6.521	−4.021	0
fd = 0.0001	−6.610	−4.026	0
fd = 0.00001	−6.503	−4.005	0

*(evaluated at $Z = [0.2, 0.4, 0]^T$)

Figure 7.   Derivative comparisons (Constraint 1).

	$\left(\dfrac{dG2}{dZ}\right)$		
	$\dfrac{dG2}{dZ_1}$	$\dfrac{dG2}{dZ_2}$	$\dfrac{dG2}{dZ_3}$
Sensitivity	3.1634	−1.4889	−10
fd = 0.1	2.6011	−1.2903	−7.829
fd = 0.01	3.1090	−1.4718	−9.849
fd = 0.001	3.1511	−1.4876	−9.983
fd = 0.0001	3.1626	−1.4746	−9.987
fd = 0.00001	3.3140	−1.4543	−10.02

*(evaluated at $Z* = [0.2, 0.4, 0]^T$)

Figure 8. Derivative comparisons (Constraint 2).

for the stiffness and damping parameters are $K = 0.365$ and $C = 0.485$. The overall process from a user's viewpoint is similar to compiling and linking a series of programs. Once the User file was prepared defining the problem, overall running of the Preprocessor, Macsyma, the Postprocessor, and the optimization took one or two minutes of clock time.

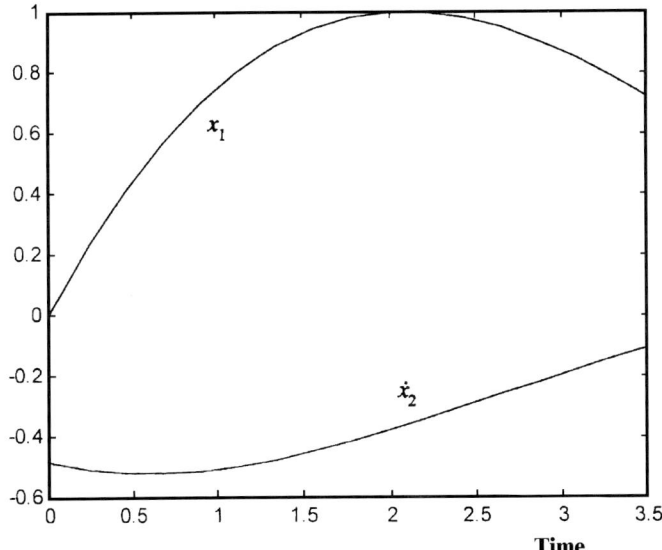

Figure 9. Optimum impact transient.

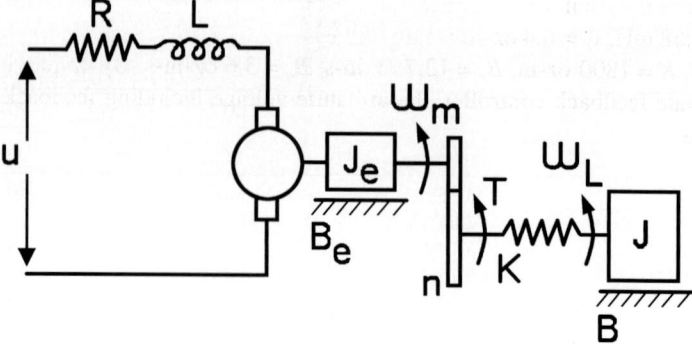

Figure 10. DC motor driving a compliant load.

2.6.2. Simultaneous Plant and Controller Design

To further explore the original Macsyma based package for automating the formulation of parameter optimization problems, the system shown in Figure 10 was considered. This system includes an armature controlled DC motor driving a flexible and lightly damped load through a gear drive. Previous study of this system (Harokopos and Mayne, 1986) had shown that matching the actuator to its load can be valuable in obtaining good closed-loop control. In this example, optimization is specifically applied to simultaneously adjust the motor-drive parameters while tuning the closed-loop control system. The state vector is

$$\underline{x} = \begin{bmatrix} i \\ \omega_m \\ T \\ \omega_L \end{bmatrix} \tag{19}$$

where i is the armature current, T is the load torque, ω_m is the motor speed and ω_L is the load speed. Input to the motor drive system is the armature voltage so that

$$\underline{\dot{x}} = \begin{bmatrix} -R/L & -K_g/L & 0 & 0 \\ K_m/J_e & -B_e/Je & -n/J_e & 0 \\ 0 & nK & 0 & -K \\ 0 & 0 & 1/J & -B_o/J \end{bmatrix} \underline{x} + \begin{bmatrix} 1 \\ 0 \\ 0 \\ 0 \end{bmatrix} u \tag{20}$$

defines the system behavior and the case is considered where $R = 8.6$ Ohms, $L = 2.58$ mH, $J_e = 0.4$ oz-in-s-s, $K_g = 0.89$ V-s, $K_m = 125$ oz-in/A, $J = 18$ oz-in-s-s, $K = 1800$ oz-in, $B_e = 13.7$ oz-in-s, $B_o = 3.6$ oz-in-s. By employing a full state feedback controller, the armature voltage including feedback becomes

$$u = z_4 \left[z_5 - \frac{z_1}{z_4} x_1 - \frac{z_2}{z_4} x_2 - \frac{z_3}{z_4} x_3 - x_4 \right] \qquad (21)$$

In this expression, z_1 through z_4 are state feedback gains while z_5 is an adjustable reference signal used to keep the load speed at a desired steady-state value during the optimization.

This problem has been explored extensively as a controls problem and sensitivity analysis optimizations have been compared to the previous optimal control studies (Narayanan, 1989). In addition, the design variable z_6 has been included in the system to bring a plant parameter into the optimization, adjusting the gear ratio n for the drive system so that the overall problem becomes

$$\text{minimize } J = \int_0^{t_f} t(\omega_L - 1)^2 \, dt$$

with

$$i \le 4$$
$$\int_0^{t_f} i^2 \, dt \le 10$$
$$0 \le z_1 \le 20 \qquad\qquad (22)$$
$$0 \le z_2 \le 20$$
$$0 \le z_3 \le 20$$
$$0 \le z_4 \le 20$$
$$0 \le z_5 \le 5$$
$$0.1 \le z_6 \le 0.5$$

where the variety of constraints are included to hold variables at realistic values and to keep the required power levels reasonable. The objective function is in the form of integral time absolute error. Transient responses for this system are shown in the load speed plots of Figure 11. Curve 1 shows the optimum transient response with a fixed gear ratio of $n = 0.27$. Curve 2 shows the optimum transient with simultaneous optimization of the gear

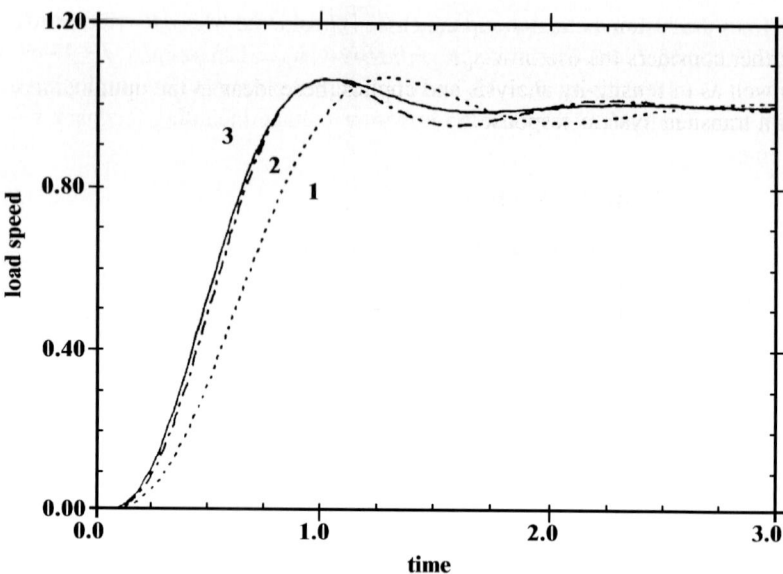

Figure 11. Optimum transients for the compliant load system.

ratio leading to $z_6 = n = 0.48$. Curve 3 in Figure 11 is the optimum transient obtained by including motor inertia as an additional design variable z_7 and with the additional constraint limits

$$0.3 \leq z_7 \leq 1.0$$

At the resulting optimum of $z_6 = 0.47$ and $z_7 = 0.3$, only modest improvement is noted by reducing the motor inertia to a lower limit.

2.7. THE USE OF "MAPLE" IN THE AUTOMIZATION PROCESS

The original work of Narayanan (1989) was based on the Macsyma symbolic manipulation package for developing sensitivity equations in the automation of the optimization process. Maple (Char *et al.*, 1992) is an attractive alternative for symbolic manipulation which has received recent attention and may be particularly of interest because of its potential for integration with Matlab (Hanselman and Littlefield, 1995) for dynamic systems studies. The work described initially in this section illustrates the use of Maple in conjunction with Matlab for a design optimization based on eigenvalues. It also illustrates the use of symbolic manipulation in the derivation of a finite element representation. The second example described, further considers the use of Maple directly in deriving equations of motion as well as in sensitivity analysis and applies these ideas in the optimization of a transient system response.

2.7.1. An Eigenvalue Problem

We consider here a simple cantilever beam with an end load and the desire to optimize the beam behavior so that it has a minimum weight while satisfying deflection, eigenvalue and geometric constraints. In specific terms we seek to

$$\text{Minimize } J = \rho \, b \, d$$
$$wrt \ b, \ d$$

$$\text{with} \tag{23}$$

$$d \leq 5b$$
$$\omega_1 \geq \omega_{min}$$
$$\Delta_s \leq \Delta_{max}$$

where b and d are the thickness and height of a beam of rectangular cross-section. The first constraint requires that the cross-section have acceptable proportions, the second assures that the lowest natural frequency in simple transverse vibration exceeds a certain minimum value and the third limits the deflection produced at the beam tip by a constant end load F.

With the finite element approach to representing the beam, transverse displacement within an element is

$$y = \sum_{i=1}^{4} \Psi_i(x) q_i(t) \tag{24}$$

where the $q_i(t)$ are generalized coordinates and the shape functions

$$\Psi_1 = 1 - 3\left(\frac{x}{h}\right)^2 + 2\left(\frac{x}{h}\right)^3$$

$$\Psi_2 = x - 2h\left(\frac{x}{h}\right)^2 + h\left(\frac{x}{h}\right)^3$$

$$\Psi_3 = 3\left(\frac{x}{h}\right)^2 - 2\left(\frac{x}{h}\right)^3 \tag{25}$$

$$\Psi_4 = -h\left(\frac{x}{h}\right)^2 + h\left(\frac{x}{h}\right)^3$$

Program to derive the mass and stiffness of a beam element

with(linalg):

```
# Shape functions (psi1,psi2,psi3,psi4)
psi1 := 1-3*((x)/h)^2+2*((x)/h)^3;
psi2 := (x)-2*h*((x)/h)^2+h*((x)/h)^3;
psi3 := 3*((x)/h)^2-2*((x)/h)^3;
psi4 := -h*((x)/h)^2+h*((x)/h)^3;

# q1d,q2d,q3d,q4d are the time derivatives of the generalized
# coordinates q1,q2,q3,q4.
# Kinetic energy of a finite element
ke := 0.5*rho*b*d*(psi1*q1d+psi2*q2d+psi3*q3d+psi4*q4d)^2;
ket := expand(ke);
kei := int(ket,x=O..h);
var := [q1d,q2d,q3d,q4d];
vv := grad(kei,var);

# Mass matrix of a finite element
M := jacobian(vv,var);

psi1d := diff(psi1,x$2);
psi2d := diff(psi2,x$2);
psi3d := diff(psi3,x$2);
psi4d := diff(psi4,x$2);
pe := 0.5*EI*(psi1d*q1+psi2d*q2+psi3d*q3+psi4d*q4)^2;
pet := expand(pe);
pei := int(pet,x=0..h);
var1 := [q1,q2,q3,q4];
dd := grad(pei,var1);

# Stiffness matrix of a finite element
K := jacobian(dd,var1);
```

Figure 12.   Maple routine for symbolic K and M matrices.

satisfy the prescribed boundary conditions for an element of length *h*. With this formulation, Maple can be conveniently used to develop the mass and stiffness matrices of individual finite elements in algebraic form. Figure 12 shows the form of the Maple program applying energy methods to obtain the matrix definitions. The resulting mass and stiffness matrices for an element are respectively

$$M^i = \rho b d h \begin{bmatrix} \dfrac{13}{35} & \dfrac{11h}{210} & \dfrac{9}{70} & \dfrac{-13h}{420} \\[2mm] \dfrac{11h}{210} & \dfrac{h^2}{105} & \dfrac{13h}{420} & \dfrac{-h^2}{140} \\[2mm] \dfrac{9}{70} & \dfrac{13h}{420} & \dfrac{13}{35} & \dfrac{-11h}{210} \\[2mm] \dfrac{-13h}{420} & \dfrac{-h^2}{140} & \dfrac{-11h}{210} & \dfrac{h^2}{105} \end{bmatrix} \tag{26}$$

and

$$K^i = \dfrac{EI}{h} \begin{bmatrix} \dfrac{12}{h^2} & \dfrac{6}{h} & \dfrac{-12}{h^2} & \dfrac{6}{h} \\[2mm] \dfrac{6}{h} & 4 & \dfrac{-6}{h} & 2 \\[2mm] \dfrac{-12}{h^2} & \dfrac{-6}{h} & \dfrac{12}{h^2} & \dfrac{-6}{h} \\[2mm] \dfrac{6}{h} & 2 & \dfrac{-6}{h} & 4 \end{bmatrix} \tag{27}$$

The elemental matrices can also be assembled by symbolic manipulation to develop a structural model. In the case of a one meter aluminum beam with three equal length finite elements, the overall mass and stiffness matrices are

$$M = bd \begin{bmatrix} \dfrac{14092}{21} & 0 & \dfrac{813}{7} & \dfrac{-3523}{378} & 0 & 0 \\[2mm] 0 & \dfrac{1084}{567} & \dfrac{3523}{378} & \dfrac{-271}{378} & 0 & 0 \\[2mm] \dfrac{813}{7} & \dfrac{3523}{378} & \dfrac{14092}{21} & 0 & \dfrac{813}{7} & \dfrac{-3523}{378} \\[2mm] \dfrac{-3523}{378} & \dfrac{-271}{378} & 0 & \dfrac{1084}{567} & \dfrac{3523}{378} & \dfrac{-271}{378} \\[2mm] 0 & 0 & \dfrac{813}{7} & \dfrac{3523}{378} & \dfrac{7046}{21} & \dfrac{-2981}{189} \\[2mm] 0 & 0 & \dfrac{-3523}{378} & \dfrac{-271}{378} & \dfrac{-2981}{189} & \dfrac{542}{567} \end{bmatrix}$$

and

$$
K = Ebd^3
\begin{bmatrix}
54 & 0 & -27 & \dfrac{9}{2} & 0 & 0 \\[2mm]
0 & 2 & \dfrac{-9}{2} & \dfrac{1}{2} & 0 & 0 \\[2mm]
-27 & \dfrac{-9}{2} & 54 & 0 & -27 & \dfrac{9}{2} \\[2mm]
\dfrac{9}{2} & \dfrac{1}{2} & 0 & 2 & \dfrac{-9}{2} & \dfrac{1}{2} \\[2mm]
0 & 0 & -27 & \dfrac{-9}{2} & 27 & \dfrac{-9}{2} \\[2mm]
0 & 0 & \dfrac{9}{2} & \dfrac{1}{2} & \dfrac{-9}{2} & 1
\end{bmatrix}
\tag{28}
$$

The resulting system equation is then

$$
M\,\underline{\ddot{y}} + K\,\underline{y} = D\,\underline{F}
$$

where the y's are nodal displacements and the influence matrix for an end load is

$$
D^T = [0 \quad 0 \quad 0 \quad 0 \quad 1 \quad 0]
$$

The character of the dynamic system can also be considered in the first order state equation form

$$
\begin{bmatrix} M & 0 \\ 0 & M \end{bmatrix}
\begin{bmatrix} \underline{\dot{y}} \\ \underline{\ddot{y}} \end{bmatrix}
=
\begin{bmatrix} 0 & M \\ -K & 0 \end{bmatrix}
\begin{bmatrix} \underline{y} \\ \underline{\dot{y}} \end{bmatrix}
\tag{29}
$$

or

$$
B\,\underline{\dot{x}} = A\,\underline{x}
$$

In this format, the eigenvalue sensitivities for the system can be determined using the convenient expression of Equation 16. The derivatives for the A and B matrices required in Equation 16 can be obtained symbolically. The other gradient information required for the optimization problem of Equation 23 can be readily obtained. Symbolic manipulation is convenient for obtaining tip deflection sensitivity in the finite element setting.

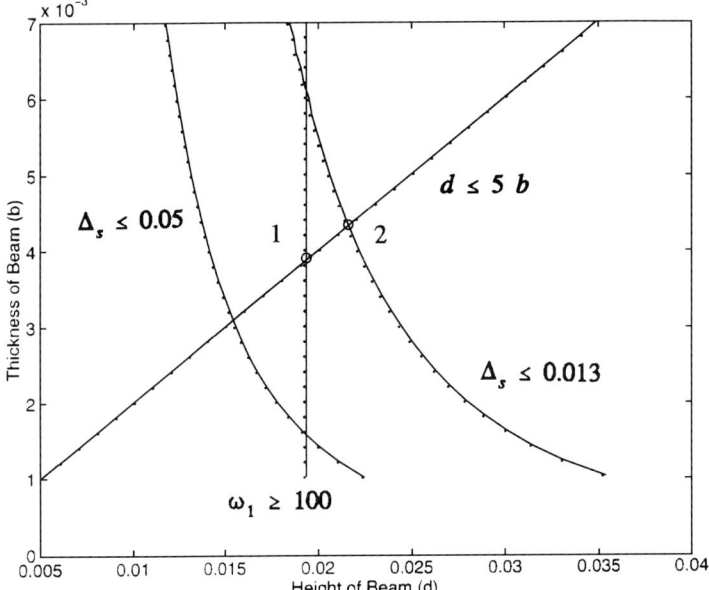

Figure 13. Constraints and optima for the beam optimization.

Matlab's sequential quadratic programming algorithm has been used to solve the optimization problem using the function and derivative files generated by Maple. The solution obtained for the case

$$E = 7.0 \; E10 \; \text{N/m}^2$$
$$\rho = 2710 \; \text{kg/m}^3$$
$$h = 1/3 \; \text{m}$$
$$F = 10 \; \text{N}$$
$$\omega_{\text{min}} = 100 \; \text{rad/sec}$$
$$\Delta_{\text{max}} = 0.05 \; \text{m}$$

shows an optimum thickness $b = 0.0039$ m and height $d = 0.0194$ m with a minimum beam mass of 0.205 kg. Figure 13 shows the constraint plots for this optimization problem. The optimum in this case is at the intersection of the natural frequency constraint with the geometric constraint shown as point 1. The deflection constraint is not active. Revising the problem with a tightened deflection constraint of $\Delta_s \leq 0.013$ results in an optimum at point 2 in Figure 13 where the deflection and cross-section geometry constraints

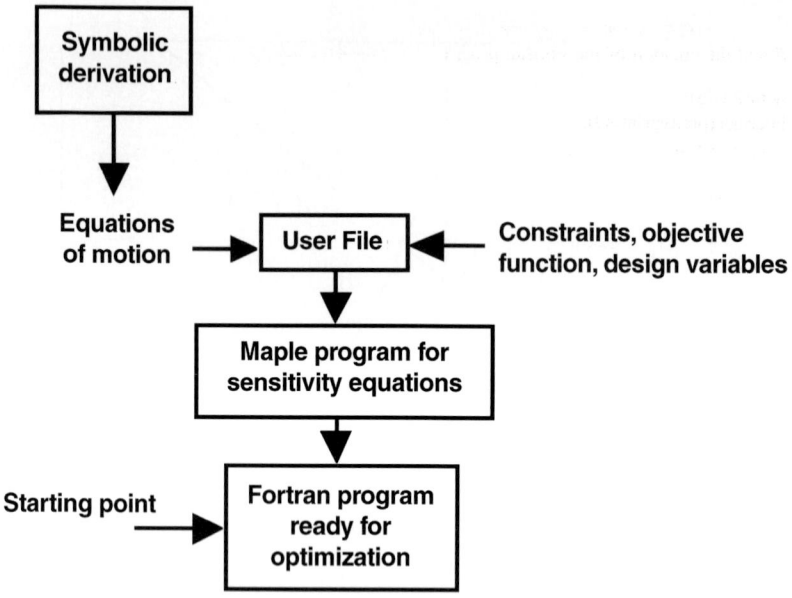

Figure 14. Ducourau's approach for automating optimization.

are active and the natural frequency constraint is inactive. This new optimum is at $b = 0.0043$ and $d = 0.0217$ where $m = 0.253$ kg.

2.7.2. Transient Optimization

The process of performing structural and dynamic transient optimization with sensitivities generated symbolically has been considered by Ducourau (1995) using Maple for symbolic operations. The environment developed by Ducourau can generate symbolic finite element representations following the pattern of Figure 12. It also generates sensitivity information with Maple and uses Maple capabilities to directly write the Fortran code required for solving the optimization problem. The approach follows many of the ideas initially described by Narayanan (1989) for linear systems and also extends the concept to nonlinear state equation forms. Figure 14 shows the overall structure of Ducourau's strategy. Figure 15 contains a major segment of the Maple sensitivity program. It begins by reading a User file describing the system and then forms the derivative expressions. The next segment of the program (not shown) generates the Fortran programming for the objective function, constraints and derivatives as well as the call to the IMSL (1991) routine NCONG which actually performs the optimization.

```
# Program for computation of the gradients
# and the creation of the Fortran program - Lazare Ducourau

with(linalg):
interface(prettyprint=O):
read 'userf.mp':

# evaluation of F
if type(A,matrix) then
F:=evalm(A&*x+B&*U):
else F:=convert(convert(F,array),matrix):
fi:
F := subs(t=time,op(F)):

# substitutions
gd := subs(t=time,op(gd )):
gds:= subs(t=time,op(gds)):

# gradients of the system
FV       :=convert(F,vector):
HSV      :=convert(convert(hs,list),vector):
GSV      :=convert(convert(gs,list),vector):
GDV      :=convert(convert(gd,list),vector):
GDSV     :=convert(convert(gds,list),vector):
DFX      :=subs(x=xx,jacobian(FV ,x)):
DHSX     :=subs(x=xtemp,jacobian(HSV ,x)):
DGSX     :=subs(x=xtemp,jacobian(GSV ,x)):
DGDX     :=subs(x=xx,jacobian(GDV ,x)):
DGDSX    :=subs(x=xx,jacobian(GDSV,x)):
DphiX    :=subs( { x=xx,abs=ab },grad(phi,x)):

for i from 1 to nz do
zd[i]:=z[i]:
od:

DFZ      :=subs(x=xx,map(grad,FV ,convert(zd,list))):
DHSZ     :=subs(x=xtemp,map(grad,HSV ,convert(zd,list))):
DGSZ     :=subs(x=xtemp,map(grad,GSV ,convert(zd,list))):
DGDZ     :=subs(x=xx,map(grad,GDV ,convert(zd,list))):
DGDSZ    :=subs(x=xx,map(grad,GDSV,convert(zd,list))):
DphiZ    :=subs(x=xx,grad(phi,convert(zd,list))):

# total number of constraints
nc := nhs+ngs+ngd+ngds:
# total number of equality constraints
nh := nhs:
# interval width
dt := evalf((tf-ti)/ndis):

# Creation of the fortran code in the file prog.f
.
.
.
```

Figure 15. Maple program for gradient determination.

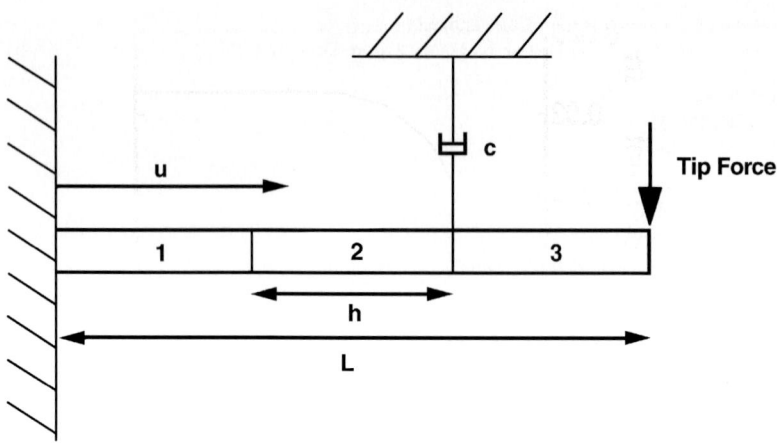

Figure 16. The beam response problem.

The system shown in Figure 16 represents one type of problem that can be addressed with Ducourau's package. A cantilevered beam is subject to a suddenly applied load F while its motion is damped by the damper C. Minimum time for the transient is desired. The dimensions of the beam are to be selected along with the damping parameter value. Constraints are applied to the beam weight and geometry. Specifically, the optimization is to

$$\text{Minimize} \quad t_f$$
$$wrt \quad b, d, C, t_f$$

with

$$\rho \, b \, d \leq \text{Mass}_{\text{max}} \qquad (30)$$
$$d \leq 5 \, b$$
$$x_{11}|_{t_f} \leq 0.001$$
$$x_{11}|_{t_f} \geq -0.001$$
$$x_5|_{t_f} \leq \Delta_{\text{max}}$$

where b is the beam width and d is the depth. The last few constraints assure that the beam tip has come to equilibrium as the final transient time is reached and that the steady state position satisfies a deflection constraint.

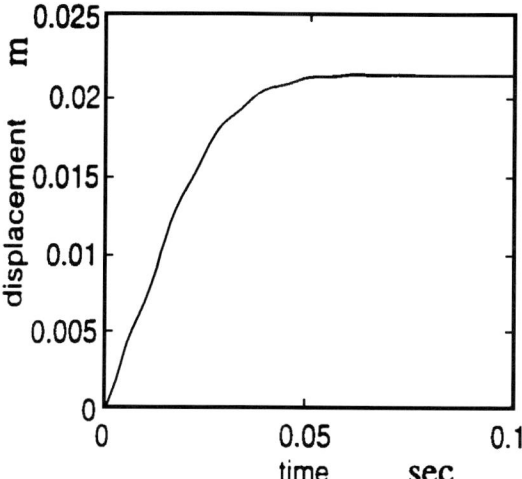

Figure 17. Optimal transient for the beam response.

As shown in Figure 16, three finite elements have been used in the beam model. Optimization results have been obtained for

$$F = 10 \text{ N}$$
$$\Delta_{max} = 0.05 \text{ m}$$
$$\text{Mass}_{max} = 0.2 \text{ kg}$$
$$\rho = 2710 \text{ kg/m}^3$$
$$E = 7.00 \ E10 \text{ N/m}^2$$

with optimum parameter values of

$$b = 0.003812 \text{ m}$$
$$d = 0.01921 \text{ m}$$
$$C = 27.91 \text{ N/m/sec}$$
$$t_f = 0.0522 \text{ sec}$$

The weight and geometry functional constraints are active while the deflection constraint is not active. The transient response of the beam tip is shown in Figure 17 for the optimum design. The beam response is well damped except for a small higher mode vibration. This vibrational mode has a node located near the absorber connection position so that it cannot be readily damped.

Ducourau has been successful in applying his strategy to several interesting problems particularly in the controls area. He considered a rotating damper member at the tip of a beam and used the damper to optimally control beam vibration. He also addressed bang-bang control action in the automated format and applied the approach to a spacecraft maneuver. The interested reader may refer to Ducourau's thesis (1995) or Ducourau *et al.* (1996).

2.8. FINAL REMARKS

The optimization of structures for dynamic performance is of increasing interest for a wide variety of applications. The state-of-the-art now allows the solution of many such optimization problems using a combination of innovative strategies and computational tools. Computer time is often a major concern in the solution of these problems with "fitting techniques" and "sensitivity analysis" as the most common methods for reducing the computational effort. Sensitivity analysis has the inherent difficulty of requiring considerable user skill and user preparation time.

This chapter has focused on symbolic manipulation as a promising tool for enhancing the use of sensitivity analysis in the optimization process. The basic ideas of sensitivity analysis useful for dynamic structural problems were discussed. Two possible approaches were described for applying symbolic manipulation in the optimization process with an emphasis on using these tools to automatically generate computer code ready for optimization. Examples were presented based on Macsyma and also on Maple. The latest results also show the possibility of symbolically generating stiffness and mass matrices for finite element representation of structural problems. It is likely that tools of this type and the use of higher level packages to automatically generate computer coding will play an increasing role in the optimal design of structural and mechanical systems.

References

1. Abadie, J. and J. Carpentier, 1969, "Generalization of the Wolfe Reduced Gradient Method to the Case of Nonlinear Constraints", *Optimization*, Academic Press.
2. Afimiwala, K.A. and R.W. Mayne, 1974, "Optimum Design of an Impact Absorber", *ASME Transactions, Journal of Engineering for Industry*, **96**(1).
3. Ashrafiuon, H. and N.K. Mani, 1990, "Analysis and Optimal Design of Spatial Mechanical Systems", *ASME Transactions, Journal of Mechanical Design*, **112**(2).

4. Borup, L. and A.R. Parkinson, 1992, "Comparison of Four Non-Derivative Optimization Methods on Two Problems Containing Heuristic and Analytical Knowledge", *Advances in Design Automation — 1992*, Hoeltzel, D.A. (Editor), American Society of Mechanical Engineers, New York.

5. Char, B.W., K.O. Geddes, G.H. Gonnet, B.L. Leong, M.B. Monagan and S.M. Watt, 1992, *First Leaves: A Tutorial Introduction to Maple V*, Springer-Verlag.

6. Corana, A., M. Marchesi, C. Martini, S. Ridella, 1987, "Minimizing Multimodal Functions of Continuous Variables with the Simulated Annealing Algorithm", *ACM Transactions Mathematical Software*, **13**.

7. Ducourau, L., 1995, "Automated Parameter Optimization for Structural and Controller Design", Master of Science Thesis, Department of Mechanical and Aerospace Engineering, State University of New York at Buffalo, Buffalo, NY.

8. Ducourau, L., T. Singh and R.W. Mayne, 1996, "Automated Parameter Optimization for Structural and Controller Design", *Proceedings of the CSME Mechanics in Design Conference*, Canadian Society of Mechanical Engineers, Toronto.

9. Goldberg, D.E., 1989, *Genetic Algorithms in Search Optimization and Machine Learning*, Addison Wesley Publishers.

10. Grandhi, R.V. and V.B. Venkayya, 1988, "Structural Optimization with Frequency Constraints", *AIAA Journal*, **26**(7).

11. Hambric, S.A., 1995, "Approximation Techniques for Broad-Band Acoustic Radiated Noise Design Optimization Problems", *ASME Transactions, Journal of Vibration and Acoustics*, **117**(1).

12. Hanselman, D. and B. Littlefield, 1995, *The Student Edition of Matlab: User's Guide*, Prentice Hall.

13. Harokopos, E.G. and R.W. Mayne, 1986, "Motor Characteristics in the Control of a Compliant Load", *AIAA Journal of Guidance, Control and Dynamics*, **9**(1).

14. Haug, E.J. and J.S. Arora, 1979, *Applied Optimal Design*, Wiley-Interscience, New York.

15. Haug, E.J., K.K. Choi and V. Komkov, 1986, *Design Sensitivity Analysis of Structural Systems*, Academic Press, Orlando.

16. Himmelblau, D.M., 1972, *Applied Nonlinear Programming*, McGraw-Hill.

17. Hooke, R. and T.A. Jeeves, 1961, "Direct Search Solution of Numerical and Statistical Problems", *Jrnl. Assoc. Comp. Mach.*, **8**.

18. IMSL Staff, 1991, *Fortran Subroutines for Mathematical Applications*, IMSL Inc.

19. Jain, P. and A.M. Agogino, 1993, "Global Optimization Using the Multistart Method", *ASME Transactions, Journal of Mechanical Design*, **115**(4).

20. Junkins, J.L. and Y. Kim, 1993, *Introduction to Dynamics and Control of Flexible Structures*, AIAA Education Series, Washington, 1993.

21. *Macsyma Vax Reference Manual*, Symbolics Incorporated, Boston, 1985.

22. Narayanan, A., 1989, "The Automation of Sensitivity Analysis and Parameter Optimization for Linear Dynamic Systems", Master of Science Thesis, Department of Mechanical and Aerospace Engineering, State University of New York at Buffalo, Buffalo, NY.

23. Narayanan, A. and R.W. Mayne, 1989, "Automating the Parameter Optimization of Dynamic Systems", *Computational Structural Mechanics and Multidisciplinary Optimization* (AD-Vol. 16), American Society of Mechanical Engineers, New York.

24. Nelder, J.A. and R. Mead, 1964, *Computer Journal*, **7**.

25. Powell, M.J.D., 1978, "A Fast Algorithm for Nonlinearly Constrained Optimization Calculations", *Proc. of the Dundee Conference on Numerical Analysis*, Springer-Verlag.

26. Reklaitis, G.V., A. Ravindran and K.M. Ragsdell, 1983, *Engineering Optimization*, Wiley Interscience.

27. Rogers, J.L. and W.J. LaMarsh, 1995, "Reducing Neural Network Training Time with Parallel Processing", NASA Technical Memorandum 110154, Langley Research Center, Hampton, Va.

29. Roozen-Kroon, P.J.M., A.J.G. Schoofs and D.H. van Campen, 1989, "Fast Numerical Shape Optimization of Bells Using Design of Experiment and Regression Techniques", *Computer Aided Optimum Design of Structures: Recent Advances*, Brebbia and Hernandez (Editors), Springer-Verlag.

30. Sandgren, E. and K.M. Ragsdell, 1980, "The Utility of Nonlinear Programming Algorithms: A Comparative Study", *ASME Transactions, Journal of Mechanical Design*, **102**.

31. Schittkowski, K., 1982, "Nonlinear Programming Codes", *Lecture Notes in Economics and Mathematical Systems*, **183** and **187**, Springer-Verlag.

32. Schuldt, S.B., 1975, "A Method of Multipliers for Mathematical Programming with Equality and Inequality Constraints", *Journal of Optimization Theory and Applications*, **17**.

33. Wang, J.H. and H.Z. Wang, 1985, "The Theory and Analysis of the Constrained Quasi-Newton Method", Mechanical Engineering Research Report, Huazhong University of Science and Technology, Wuhan, China.

3 TECHNIQUES FOR AUTOMATIC MESH GENERATION FOR SOLID MODELS

EDWIN BOENDER, WILLEM F. BRONSVOORT and MARK F. HERMELING

Faculty of Technical Mathematics and Informatics,
Delft University of Technology, Julianalaan 132, 2628 BL Delft,
The Netherlands

3.1. INTRODUCTION

Computer-Aided Design (CAD) is an important area in industrial automation, concerned with computer support of the design of a product. Several activities can be recognized that are common to the designer's job in many industries. The first is specification, where criteria for the performance of the product are laid down. The second is synthesis, where a design is created that possibly satisfies the criteria laid down during specification. The third is evaluation, the activity best suited to computer support. Evaluation can again be divided into three steps. The first step is concerned with the representation of the design of a product. This can, for example, be a drawing, a mathematical model, a geometric model, or a physical model. The second step is concerned with the quantitative analysis of the proposed design. This involves simulation to predict the properties and behaviour of the proposed design. The third step involves comparing the specification

Correspondence: Willem F. Bronsvoort, Faculty of Technical Mathematics and Informatics, Delft University of Technology, Julianalaan 132, 2628 BL Delft, The Netherlands. Phone: +31 152782533, Fax: +31 152787141, Email: bronsvoort@twi.tudelft.nl

with the results of the analysis. This may result in changing the proposed design, but it may also require a change in the specifications.

During the evaluation phase, the proposed design is analysed to determine how it will function in its working environment. An object may, for example, be subjected to stress, heat or vibrations. If analysis shows unsatisfactory behaviour under these conditions, the design has to be revised. The finite element method (FEM) is the most important tool for such testing of proposed designs.

An FE method computes an approximate solution for a physical problem, given the governing equation for the problem, a computational domain over which the variable has to be computed, and a sufficient number of boundary conditions. An example of a computational domain is the boundary of, and volume occupied by, a mechanical part. An example of a boundary condition is the load applied to the mechanical part to analyze its behaviour under stress. Boundary conditions on the computational domain can be used to specify the value of the model variable at the boundary of the domain. Mathematically speaking, they have to be imposed to obtain a unique solution to the problem. Examples of books dealing with mathematical fundamentals and applications of FE methods are.[1,2]

FE methods were originally developed within the aircraft industry for the analysis of mechanical components. The first software packages were specifically designed for calculation of displacements and stresses within mechanical components. FE methods can, however, also be applied to similar problems (involving similar mathematical formulations) in other application fields. Some of these fields are fluid dynamics, biomechanics, chemistry and acoustics.

Essential for the FE method is the decomposition of the computational domain into a number of (finite) elements. Depending on the dimension of the computational domain, elements can either be 1D, 2D, or 3D, and can be embedded within a space of the same or higher dimension. Here both 2D and 3D techniques are discussed. Examples of elements are triangles and quadrangles for a 2D domain, and elements with four, five or six faces for a 3D domain. On the boundary of the element, and possibly also in the interior, a number of nodes are defined. Figure 1 gives an example of a 2D domain represented with both triangular and quadrilateral elements. Nodes are only defined on the vertices of the elements. Here only triangular and tetrahedral elements are considered, because these are most suitable for applications in CAD.

The nodes of the FE mesh represent a discretization of the computational domain. Within this discretization, the elements of the FE mesh are supposed to be connected only at their common nodes. A solution to the problem is computed for nodal values representing the model variables at the nodes of

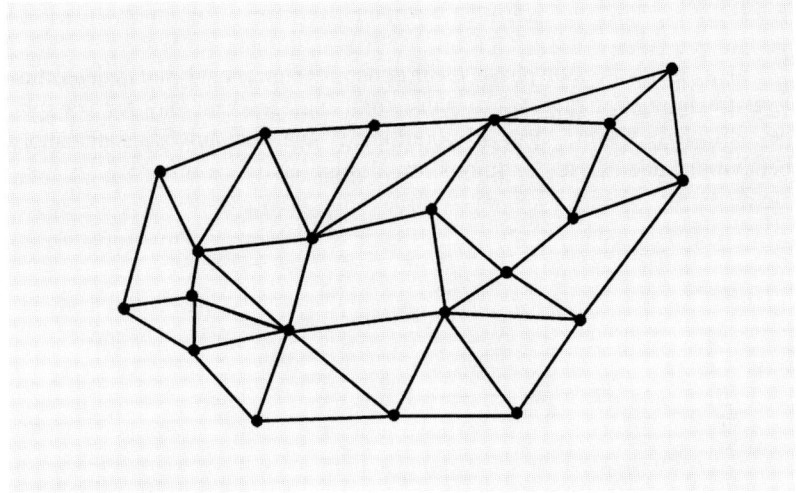

Figure 1. A 2D domain with triangular and quadrilateral elements.

the mesh. The values within the elements are then obtained by interpolation of the computed nodal values. The interpolation functions used for this are usually piecewise polynomials. These functions are zero over most of the discretized domain, and are non-zero only within a neighbourhood of a particular node. The meaning of the nodal values and the interpolation functions depends on the application: in structural analysis, the variable would be stress or displacement distance, but in heat studies, the variable would be temperature.

The variables within the physical model are usually assumed to have a certain degree of continuity over the computational domain. Because of the discretization of the domain and the use of the interpolation functions, the computed approximation for the problem will generally not have the same degree of continuity. The computed solution will be continuous and smooth within the elements of the mesh, but this is not necessarily the case at the common element boundaries.

When an FE mesh has been derived for the computational domain, the element properties can be used to formulate a local linear system of equations for each element in the mesh.[3] The unknowns in each local system of equations are related to the mesh nodes for which the interpolation functions have non-zero value. Thus, only a fraction of the total number of mesh nodes are involved in the formulation of the linear system for each element. The element properties that define the formulation of the corresponding local linear system are the interpolation functions within the element, and the geometry of the element, i.e. the position of the nodes.

Local systems share unknowns at nodes for which the interpolation function takes on a non-zero value over the elements corresponding with the local systems. The local linear systems are collected to formulate a linear system of equations for the whole mesh, involving the variables at all nodes in the FE mesh. This results in a large, sparse, system of linear equations, involving perhaps hundreds of thousands of unknowns. This linear system is then solved.

It follows from the above that the quality of the obtained solution at least partially depends on the type, shape and size of finite elements. Better approximations of the required solution can be obtained by the use of higher-order interpolation functions (involving the introduction of extra nodes within an element), and also by the use of more elements. Furthermore, the quality of the solution can be expected to degrade if the mesh contains badly-shaped elements, i.e. elements that are deformed and irregular. In 2D, for instance, elements with large obtuse angles are considered to be badly shaped.

The creation of an FE mesh, requiring the subdivision of the computational domain into elements, is a major bottleneck in the application of FE methods. The elements of an FE mesh, and their connectivities, may be explicitly specified by the user of an FE program. However, current applications often involve computational domains with complex 3D geometry. To obtain an accurate solution for such a domain, a mesh may be required consisting of tens of thousands of elements. The specification of the elements, and their connectivities, for such a domain is a tedious and error-prone task, and may require several man-months.

An alternative to the above is to use methods for semi-automatic or fully automatic generation of an FE mesh. These methods derive a mesh from a geometric model, representing the computational domain, and the boundary conditions. The geometric model mostly contains the vertices, edges and faces of the computational domain. Curved edges and faces are often approximated by planar ones, which leads to a polygonal approximation of the original object. However, curved edges and faces can also be represented exactly; an example of this is given in.[4]

Semi-automatic methods still require the user to manually subdivide the domain into a number of 'well-shaped' super-elements. These super-elements are then automatically further subdivided into smaller elements. The disadvantage of semi-automatic methods is that creation of the super-elements may be difficult because of the complex geometry of the domain.

The alternative is fully automatic generation of an FE mesh, given a geometric model representing the computational domain, the boundary conditions, and some meshing constraints involving, for example, the required element size. A density function is used to control the element size. The density function is inversely proportional to the required element size. Fully automatic generation requires more complex meshing procedures, but is the more attractive alternative, since it minimizes user involvement in the discretization of the domain. This approach will therefore be pursued here.

The second section discusses the requirements for an FE mesh generator. Mesh generation with separate node and element generation is discussed in the third section; the mesh is generated by first creating all nodes, and then connecting these nodes to form the elements of the mesh. The fourth section describes approaches that generate nodes and elements simultaneously. It is also possible to subdivide the object into simpler pieces that can be meshed more easily; several approaches using such strategy are discussed in the fifth section. The next two sections describe the use of two alternative models to the 'standard' geometric model for use in mesh generation. In the last but one section, methods are described to improve the elements of the mesh by changing the geometry and/or the topology of the mesh. In the last section, an evaluation of the different approaches is given.

3.2. REQUIREMENTS FOR FINITE ELEMENT MESH GENERATORS

The requirements for an FE mesh generator partially depend on the application environment. To be suitable for practical applications in structural analysis, automatic mesh generators for both 2D and 3D models should satisfy the following requirements:

1. The mesh generator must be able to cope with an object having an arbitrary shape, e.g. holes or voids in its interior should be possible.
2. The user of an FE mesh generator must have adequate control over the mesh density in different regions of the object to be meshed. This control is necessary since the user may already know from experience whichparts of the model require a denser mesh.
3. The mesh generator must generate a mesh that is topologically and geometrically correct.
4. The mesh generator must generate a mesh of the highest quality possible, e.g. the mesh should contain as few badly-shaped elements as possible.
5. The mesh generator must generate a mesh that matches the geometric model as good as possible; boundary nodes of the mesh should be positioned exactly on the edges and faces of the model.

6.   The mesh generator should require a minimal amount of user input.
7.   Generation of a mesh should require a minimal amount of computation
     time. This is a less important requirement, since the analysis phase
     usually requires much more computation time than the mesh generation
     phase.

Regarding requirement four, it should be pointed out that the quality of
a mesh strongly influences well the results of the analysis agree with the
exact solution. The quality of the mesh depends on its density and the shape
of the elements. For instance, it is well known that for triangular elements
obtuse angles will degrade the accuracy of the results, whereas acute angles
do not.[5] Thus, as few triangles with obtuse angles as possible should be
created. Furthermore, the mesh density should be higher where the gradient
of the function being approximated is higher, because the FE method computes the properties of an approximation to the physical model that does not
have the same order of continuity as the model, implying that the computed
result will differ more from the exact solution in regions where the gradient
is higher. One approach to solve this problem is to use local error estimators
during the analysis, which indicate where a mesh has to be refined and where
coarser elements can be used.

The fifth requirement has been included to ensure that the geometry is
not simplified too much when deriving a mesh from a model. It has been
shown that even minor simplifications of the geometry can lead to large
errors.[6]

It follows from the above that the requirements for a 'good' mesh are
difficult to quantify. The quality of a mesh depends on the shape of the
elements, their size, and the gradient of the exact solution. There is not one
clear criterion, instead there is a set of vague 'conditions'. It is, for example,
clear that obtuse triangles give a less accurate result, but when to classify
a triangle as too badly shaped is less clear, since the accuracy of the solution
is also dependent on the gradient within that triangle.

3.3. MESH GENERATION USING SEPARATE NODE AND
ELEMENT GENERATION

In the methods of this type, the mesh generation is divided into two separate
phases:

–   First, nodes are generated on the boundary of the object and inside the
     object.
–   Second, the generated nodes are combined into elements with three
     edges in 2D or four faces in 3D.

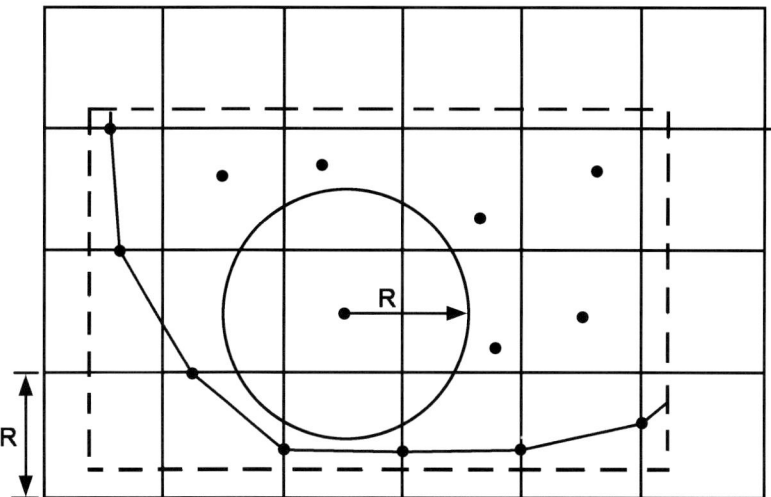

Figure 2.   Random generation of internal nodes inside a zone (dashed lines).

Triangulation and tetrahedrization methods can only form elements that are well-shaped if the node distribution allows this. It thus follows that a good node generation method is essential to obtain well-shaped elements. Several node generation strategies are discussed below.

3.3.1.   Node Generation Strategies

The method of Cavendish[7] generates nodes for a 2D object. The user divides the object space into a number of zones. For each zone a different required mesh density is defined. A square grid is then placed over each zone. The size of the squares is determined by the required mesh density for that zone.

A node is randomly positioned inside each square. It has to satisfy the following conditions (see Figure 2):

– The distance of the node to all previously generated nodes must be larger than the size of the square.
– The node must lie both inside the boundary of the object and inside the zone under consideration.

If these conditions are not satisfied, a new node is generated at random. If after five attempts still no acceptable node has been found, no node is generated.

There are two disadvantages to this method:

- A number of geometrical checks must be performed for each node to determine whether it can be accepted.
- The random positioning of nodes can cause badly-shaped elements. Moreover, a square can be left without a node after five attempts, whereas an acceptable node is possible.

Lo[8] describes a more effective 2D method for the generation of nodes. Again, the user divides the object space into a number of zones. Each zone has a different required mesh density. First, horizontal lines are taken through the object. The distance between these lines is determined by the required mesh density. Then nodes are generated along these lines. The distance between these nodes is again determined by the required mesh density. The method can easily be extended to 3D by intersecting an object with a grid of parallel lines. The parallel lines can be processed by the 2D algorithm.

More recent methods are physically-based, such as the method described in.[9] These methods are interesting because they are based on natural phenomena, and so can result in good meshes.

The physically-based 3D method in[9] was inspired by the observation of floating soap bubbles in liquid. Connecting bubble center points, results in a mesh of triangles, most of which are almost equilateral. So a model that simulates the physics of soap bubbles is used for 2D mesh generation. This idea can easily be extended to 3D by simulating soap bubbles in a volume.

A node is seen as a bubble with a point mass. The bubbles are affected by an inter-bubble force field, i.e. they repel and attract each other. An equilibrium configuration is computed by solving the governing equations of motion numerically. This results in a static force balance.

The user controls the mesh density by the definition of a density function, which simulates a temperature field. The temperature distribution causes bubbles in warm areas to grow, and bubbles in cold areas to shrink. Two adjacent bubbles are tangent to each other in the final configuration. So the mesh density corresponds to the bubble diameters.

Nodes for the edges, faces and the volume are generated in separate stages. A bubble can be classified into one of the following four types: vertex-bubble, edge-bubble, face-bubble or volume-bubble (see Figure 3).

Nodes are generated in the following four steps:

- Generate bubbles on the boundary.
- Compute the number of bubbles that will fit in the interior.
- Generate bubbles in the interior.
- Solve the equations of motion for the bubble configuration.

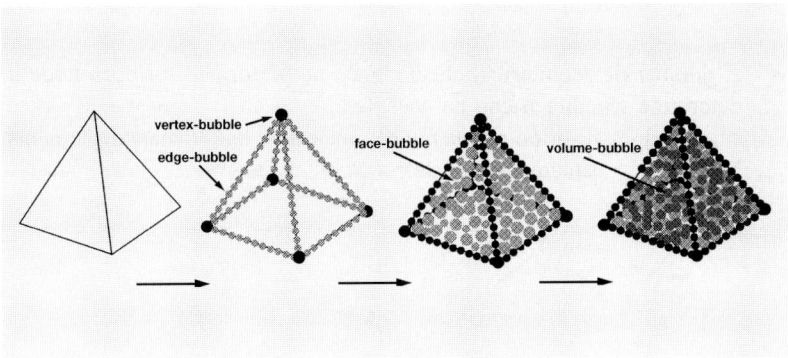

Figure 3.   Bubble types and generation order.

To simplify the model, the non-linear inter-bubble forces are approximated by linear equations.

This method results in a reasonable element shape regularity, so few badly-shaped elements will be formed. However, it is not the most efficient approach, which is due to the initial intent of Shimada and Gossard[9] to make the model as realistic as possible.

3.3.2. Conclusions

The methods of Cavendish[7] and Lo[8] give good results in 2D, but less good results in 3D. In 3D, often geometric details are lost.

The physically-based methods are well suited for use in both 2D and 3D. The physically-based methods mimic natural shape finding processes. These methods consist of the following three steps:

–   Determine the physical model.
–   Define the governing equations.
–   Compute a numerical solution for the equations.

Within these steps, there is a lot of freedom to select what level of detail will be used. The model, the governing equation, and the solution used in[9] are very realistic. This causes the method to be less efficient as possible.

3.3.3. Delaunay Triangulation

There are many different 2D and 3D methods that create triangular or tetrahedral elements by combining previously created nodes. Most of these methods are based on Delaunay triangulation/tetrahedrization.

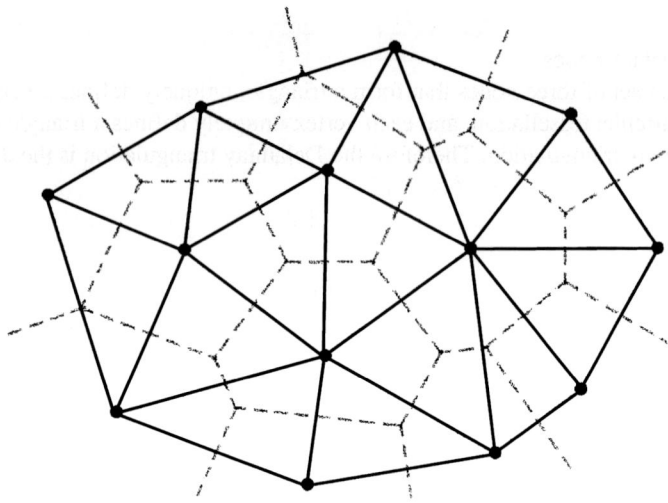

Figure 4. Voronoi tessellation (dashed lines) with corresponding Delaunay triangulation (solid lines) for a set of nodes.

The Delaunay triangulation is seen as the best triangulation of a model. It maximizes the sum of the smallest angles over all triangles in the triangulation.[10,11] Elements with sharp angles are thus avoided as much as possible. Another advantage of the Delaunay triangulation method is that, unlike many other methods, it is based on a mathematical theory.

The Delaunay triangulation is first defined for 2D, and then extended to 3D.

For a set of nodes, polygons can be defined that contain those points of the plane that are closer to a particular node than to any other node. This set of polygons is called the Dirichlet or Voronoi tessellation. By connecting all nodes for which the corresponding polygons share an edge, triangles are formed. This set of triangles forms the convex hull of all the nodes, and is called the Delaunay triangulation (Figure 4).

At least three polygons meet in each vertex of the Dirichlet tessellation. If three polygons meet in a vertex, then that vertex of the tessellation is equidistant to the three nodes of these polygons. Each vertex of the Dirichlet tessellation is thus the center of the circumscribing circle of the triangle formed by the three nodes. An important property of such a Delaunay triangle is that the circumscribing circle does not contain any other node. If more than three polygons meet in a vertex of the tessellation, then a unique Delaunay triangulation cannot be formed, because there are at least four nodes equidistant to that vertex. In this case, a (non-unique) Delaunay triangulation can be obtained by triangulation of the convex hull of the equidistant nodes.

Each set of three nodes that form a triangle, uniquely defines a vertex of the Dirichlet tessellation, and each vertex uniquely defines a triangle of the Delaunay triangulation. Therefore the Delaunay triangulation is the dual of the Dirichlet or Voronoi tessellation.

The above definitions can be extended to 3D as follows:

– The Dirichlet tessellation will now be formed by a set of polyhedra defined by several nodes, such that all points inside a polyhedron are closer to the associated node than to any other node.
– Normally, four polyhedra meet in each vertex of the tessellation. If more then four polyhedra meet in a vertex of the tessellation, then a unique Delaunay tetrahedrization cannot be formed.
– The Delaunay triangulation is formed by connecting the nodes for which the corresponding Voronoi polyhedra share a face.
– Each vertex of the Dirichlet tessellation is the center of the circumscribing sphere of the tetrahedron defined by the four nodes belonging to the Voronoi polyhedra sharing that vertex.
– There are no other nodes inside the circumscribing sphere of a tetrahedron than those of the tetrahedron itself.

3.3.4. Watson's Algorithm

Watson's algorithm[12] for the Delaunay triangulation of a 2D node set first creates an initial triangle such that all nodes fall within this triangle. The nodes are then inserted one after the other in the following steps:

– It is determined which triangles have the node within their circumscribing circle. These triangle are marked for removal.
– An insertion polygon is constructed using the marked triangles. All edges that occur twice are removed, because these edges do not belong to the boundary of the insertion polygon (see Figure 5). The node to be inserted always lies inside the insertion polygon.
– New triangles are formed by combining every edge of the insertion polygon with the inserted node. This is done by connecting the vertices of the edges to the node. The newly formed edges lie inside the insertion polygon (see Figure 6).

Watson[12] proves that the Delaunay properties of the triangulation are preserved with every insertion of a node. An exception occurs if there are more than three Voronoi regions of the tessellation sharing the same vertex.

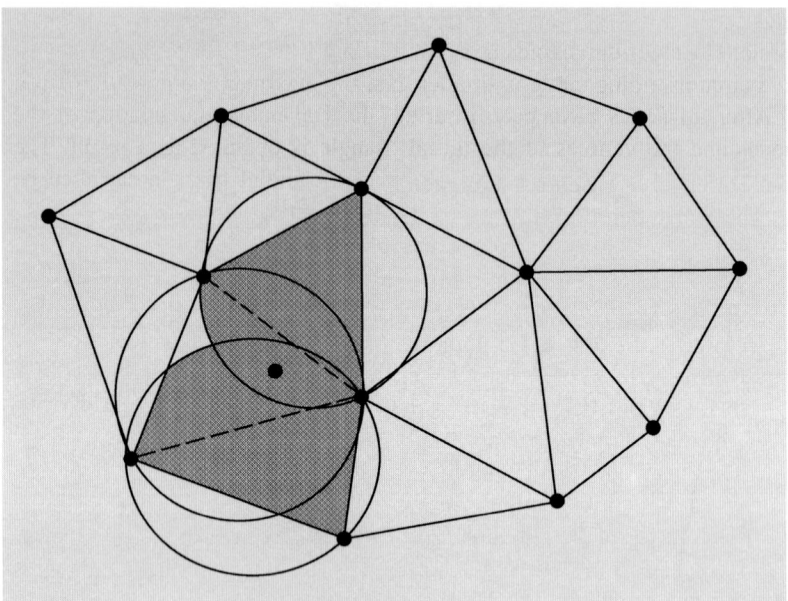

Figure 5.   Insertion of a new node to a 2D Delaunay triangulation using Watson's algorithm.

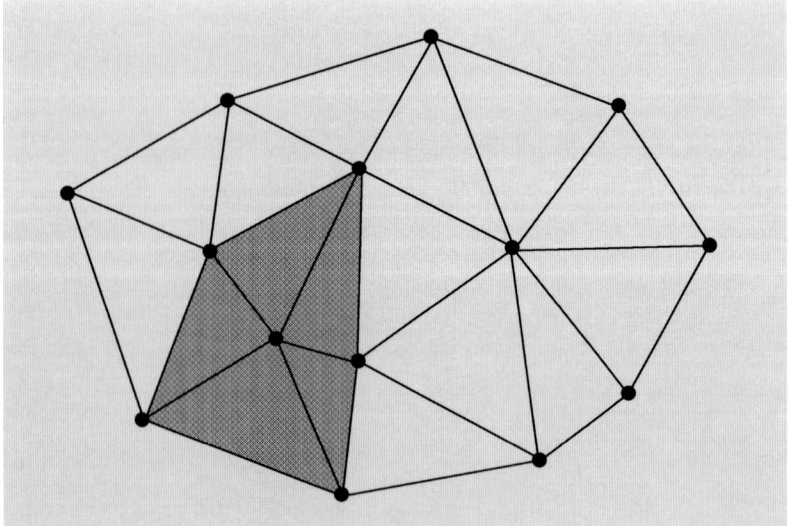

Figure 6.   The new 2D Delaunay triangulation.

This happens when inserting a node that is too close to a circumscribing circle. The algorithm avoids this problem by ignoring a node if it's distance to a circumscribing circle is smaller than the maximum computation error.

After all nodes have been inserted, the Delaunay triangulation of the nodes and the vertices of the initial triangle is obtained as a result. The triangles that have a node at a vertex of the initial triangle can then be removed.

The above algorithm can be extended to 3D as follows:

- The algorithm first creates an initial tetrahedron such that all nodes fall within this tetrahedron. After tetrahedrization, all tetrahedrons that have a node at a vertex of the initial tetrahedron have to be removed.
- For each node to be inserted, it is determined which tetrahedrons have the node within their circumscribing sphere. These tetrahedrons are marked for removal, and an insertion polyhedron is constructed.
- New tetrahedrons are formed by connecting the vertices of the triangles of the insertion polyhedron to the inserted node.

3.3.5. A Delaunay Algorithm for Triangulating Polygons and Polyhedra

When using Watson's algorithm, the triangulation domain is considered to be equal to the convex hull of the node set. This implies that, if this algorithm is used to triangulate the vertices of a concave polygon or polyhedron, triangles (2D) or tetrahedra (3D) may be formed that intersect or lie outside the object boundary.

De Floriani et al.[13] use the concept of a constrained Delaunay triangulation, which prevents that triangles (tetrahedra) are formed that intersect or lie outside the boundary of the object. This requires the concept of visibility of nodes. A node A is visible from a node B if the line AB lies completely inside, or on the boundary of the object.

A definition of a constrained Delaunay triangulation that can cope with an object having an arbitrary shape, defines a triangle ABC as a Delaunay triangle if there exists no node D such that D is within the circumscribing circle of ABC and D is visible from nodes A, B and C (see Figure 7).

3.3.6. Cavendish et al.'s Method

Cavendish et al.[14] describe a 3D mesh generator that uses an adapted version of Watson's algorithm for 3D. One adaptation is the handling of nodes that lie too close to the boundary of a circumscribing sphere of a tetrahedron. Watson discards such node during tetrahedrization, but Cavendish et al.

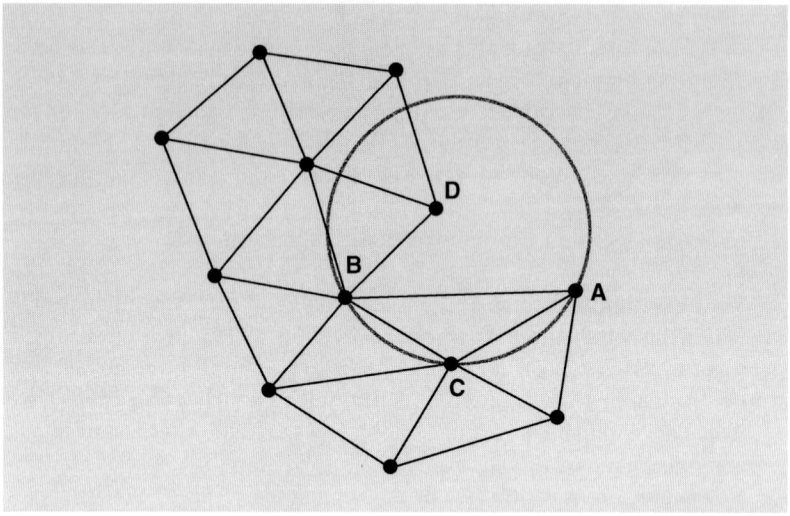

Figure 7. Constrained Delaunay triangulation. Triangle ABC is a Delaunay triangle, because, although node D is inside the circumscribing circle of ABC, D is not visible from A and C.

move such a node over a short distance so that it does not form an exception. The node is then inserted to the tetrahedrization, and moved back to its original position afterwards.

Another adaptation has been made for the occurrence of slivers. Slivers are flat, badly-shaped, tetrahedrons with a small volume (see Figure 8). Slivers must be detected and removed.

To remove slivers, the tetrahedrons that share an edge with this tetrahedron have to be taken into consideration. Considering a sliver ABCD, two

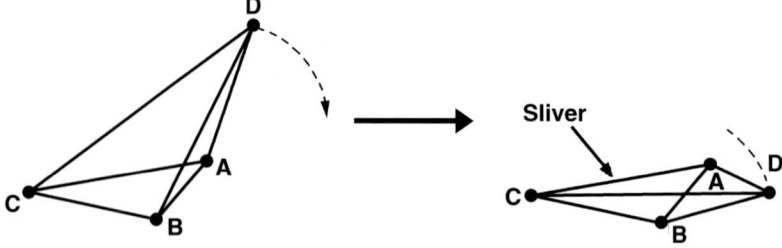

Figure 8. Delaunay tetrahedron deformed into a sliver.

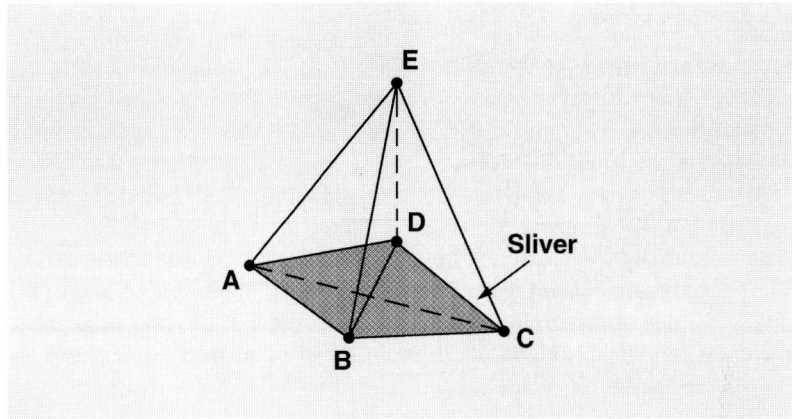

Figure 9. A removable sliver.

cases can be distinguished (see Figure 9 and Figure 10). The first case is that of two adjacent tetrahedrons sharing a node, say E (Figure 9). Here the sliver is removed and the tetrahedrons {ABDE, BCDE} are replaced by {ABCE, ACDE}. In the second case there are no two tetrahedrons that share a node (Figure 10). The position B will be moved to, for example, the node (B+E)/2. The shape of element ABCD is thus slightly improved.

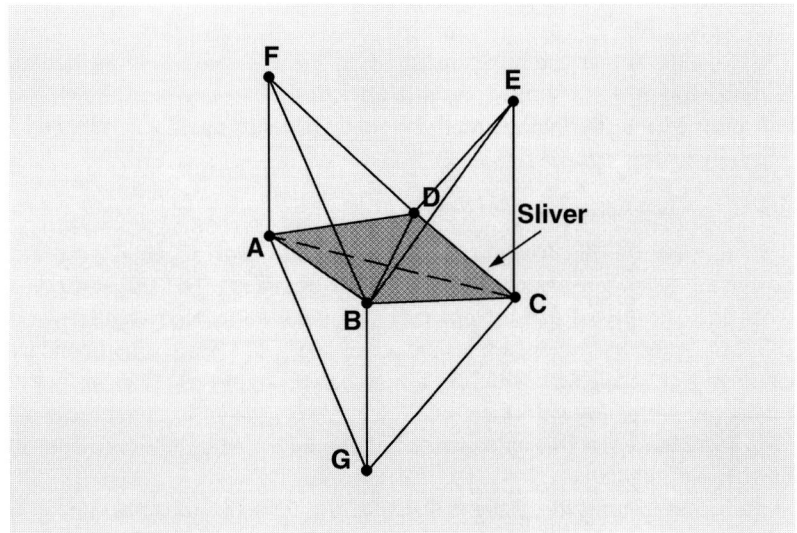

Figure 10. An unremovable sliver.

3.3.7. Conclusions

Most 2D and 3D methods that create triangular or tetrahedral elements by combining previously created nodes produce a Delaunay triangulation/ tetrahedrization. This maximizes the sum of the smallest angles over all triangles in the triangulation. Unlike many other methods, it is based on a mathematical theory. The Delaunay triangulation for 2D usually gives good meshes. For 3D, however, distorted elements may still be created.

In[12] an algorithm is described that is based on a Delaunay triangulation. The triangulation domain is considered to be equal to the convex hull of the nodes. So only convex objects can be triangulated. If concave objects are triangulated with this algorithm, elements will be formed that intersect or lie outside the object boundary.

De Floriani *et al.*[13] describe an adapted version of the Delaunay triangu- lation, which prevents that elements that intersect or lie outside the boundary of the object are formed.

Cavendish *et al.*[14] describes a 3D mesh generator that is based on Watson's algorithm. Their method can handle the occurrence of badly-shaped tetra- hedrons with a small volume. These are detected and removed. However, the detection criterion is based on a rule of thumb, so it is not guaranteed that all slivers will be detected.

3.4. MESH GENERATION USING SIMULTANEOUS NODE AND ELEMENT GENERATION

The methods that use this principle start by generating a triangulation/ tetrahedrization for a limited number of nodes. Then nodes are generated and inserted into the object, until the required mesh density is obtained.

3.4.1. Advancing Front Method

This method for the triangulation/tetrahedrization of an object uses an advancing front consisting of edges (2D) or triangles (3D). The advancing front encloses the part of the object that has not yet been triangulated. At the start of the mesh generation, the advancing front lies on the boundary of the object. Using an edge (2D) or a triangle (3D) of the advancing front, a new mesh element is formed, and the advancing front is adapted. So the advancing front encloses an increasingly smaller area, until the entire object has been triangulated.

To locally control the shape and size of the elements, a background grid of triangles (2D) or tetrahedrons (3D) is used. The following meshing parameters have to be specified at every node of a 2D background grid:[15]

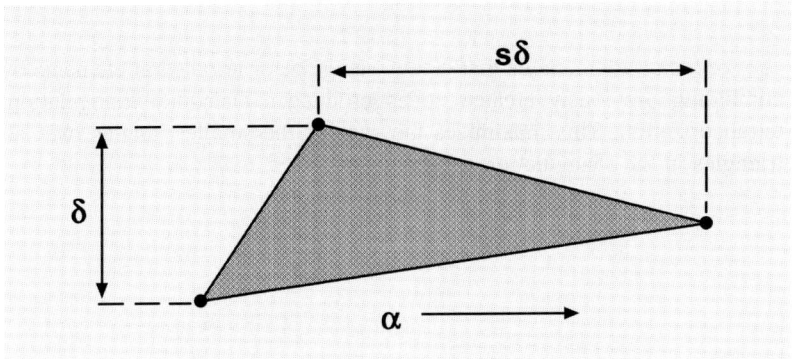

Figure 11. Definition of background grid parameters in 2D.

- δ, the required distance between nodes.
- s, the extent to which an element should be stretched.
- α, the direction in which the elements should be stretched.

A generated element will be about $s\delta$ long in the direction parallel to α, and δ long in the direction perpendicular to α (Figure 11).

In 3D, five parameters have to be specified for every node.[15] These are the required distance between nodes δ, two scaling factors s_1 and s_2, and their directional vectors α_1 en α_2 (Figure 12).

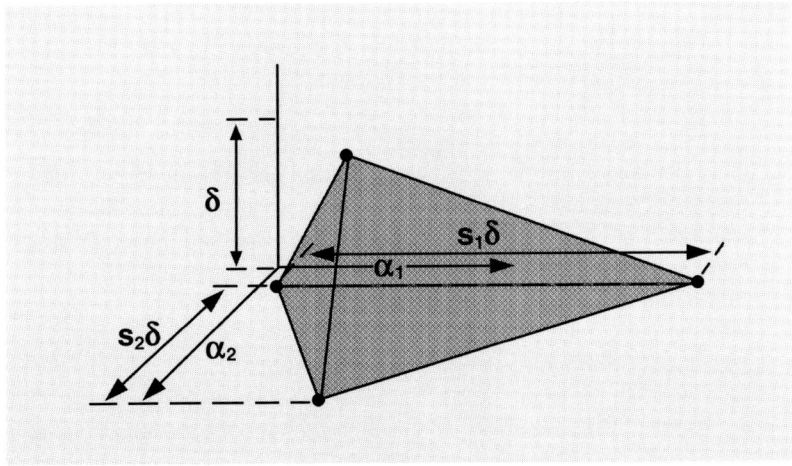

Figure 12. Definition of background grid parameters in 3D.

E. BOENDER *et al.*

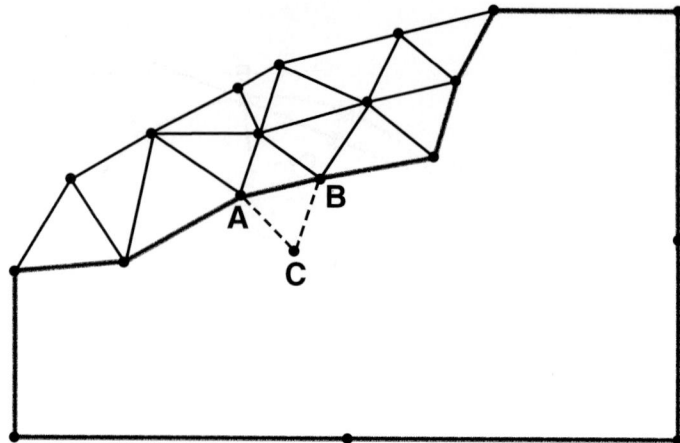

Figure 13. Selecting edge AB in the advancing front.

During mesh generation, the values of the parameters at an arbitrary position are obtained by interpolating their values at vertices of the background grid. The background grid must cover the entire object. Specification of the elements of the background grid and the parameters in every node must be done by hand. This is not a serious problem, because the background grid normally consists of a limited number of elements.

With the background grid, the density and shape of the elements can be varied along the object. This is very important when solving flow problems, where this advancing front method is often used. The ability to vary the shape of elements is less important for stress calculations in mechanical engineering. Here a single scaling factor is mostly sufficient.

The 2D algorithm consists of the following steps:[15,16]

- Define a background grid consisting of triangles.
- Generate nodes on the boundary of the object. The required distance between the nodes on the boundary is obtained from the values of δ, s and α. The edges on the boundary of the object and the edges on the boundary of holes in the object form the initial advancing front.
- Select the next edge in the advancing front. This will be the shortest edge of the front, say AB (see Figure 13).
- Execute the following steps for this edge AB:
    - Calculate the value of δ_M, s_M and α_M for the centre M of AB. This is done by determining inside which triangle of the background grid M lies. Next the parameters of M are determined by interpolation of the values of δ, s and α at the vertices of the triangle.

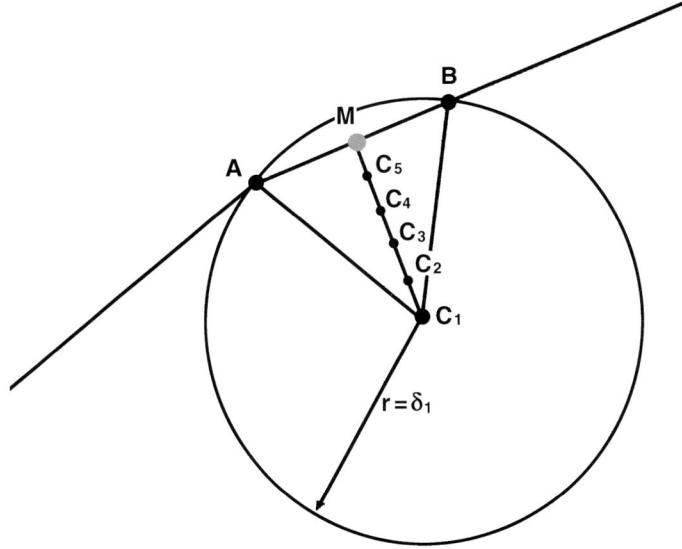

Figure 14. Potential new node computed for edge AB of the advancing front.

– Determine the ideal node C_1 that has a distance δ_1 to A and B. Assuming $s_M = 1$, so theelements are not stretched, then δ_1 is calculated as follows:

$$\delta_1 = 0.55\ AB \qquad \text{if } \delta_M < 0.55\ AB \qquad (1)$$

$$\delta_1 = \delta_M \qquad \text{if } 0.55\ AB < \delta_M < 2\ AB \qquad (2)$$

$$\delta_1 = 2\ AB \qquad \text{if } 2\ AB < \delta_M \qquad (3)$$

These values are chosen to prevent that extremely badly-shaped elements are formed, and are based on experience.
– Determine which already existing nodes lie within the circle with centre C_1 and radius δ_1. Put these nodes in order of increasing distance from C_1 into a list. Add C_1, and the nodes $C_2 \ldots C_5$ that lie on the line MC_1, to the end of the list (see Figure 14).
– Find the first node N_j in the list for which:
  – the edges AN_j and N_jB do not intersect any edge of the advancing front.
– Form the new triangle with the edge AB and the node N_j. If N_j did not already exist (i.e. is any C_i), then a new node has been generated.

- Update the advancing front. Remove the edges that can no longer be used to form any triangles, and add the newly formed edges to the advancing front.

The mesh generation process continues to select edges from the advancing front until the entire object has been triangulated.

The 3D algorithm consists of the following steps:[16,17]

- Define a background grid consisting of tetrahedrons.
- Generate nodes on the boundary of the object. The required distance between nodes on the boundary is obtained from the values of δ, s_1, s_2, α_1 and α_2.
- Triangulate the boundary of the object by using the 2D method. To do this, projections of the boundary from 3D to 2D and back are needed.[17] The obtained triangulation of the boundary forms the initial advancing front of the tetrahedrization.
- The remainder of the algorithm is a straightforward generalization of the algorithm for 2D, with triangle instead of edge and tetrahedron instead of triangle.

3.4.2. Delaunay Triangulation Using a Boundary Representation

Frey[18] describes a method that exploits the facts that nodes are inserted sequentially to a Delaunay triangulation and that after each insertion the triangulation retains its Delaunay properties. Consequently, nodes do not have to be generated all at once, but each node can be generated just before inserting it into the current triangulation. It is thus possible to take the current triangulation into account when creating the next node.

The mesh density of the 2D algorithm is controlled by a density function $d(x,y)$. The function $d(x,y)$ specifies the required distance to the nearest node for each node with coordinates (x,y).

The 2D algorithm consists of the following steps:

- Generate nodes on the boundary of the object complying with the density function.
- Generate the Delaunay triangulation for the nodes on the boundary of the object.
- Insert nodes into the interior of the object and update the triangulation. This is done as follows:
  - Select a triangle that has the center of its circumscribing circle inside the triangle, thus obtuse triangles are not selected.

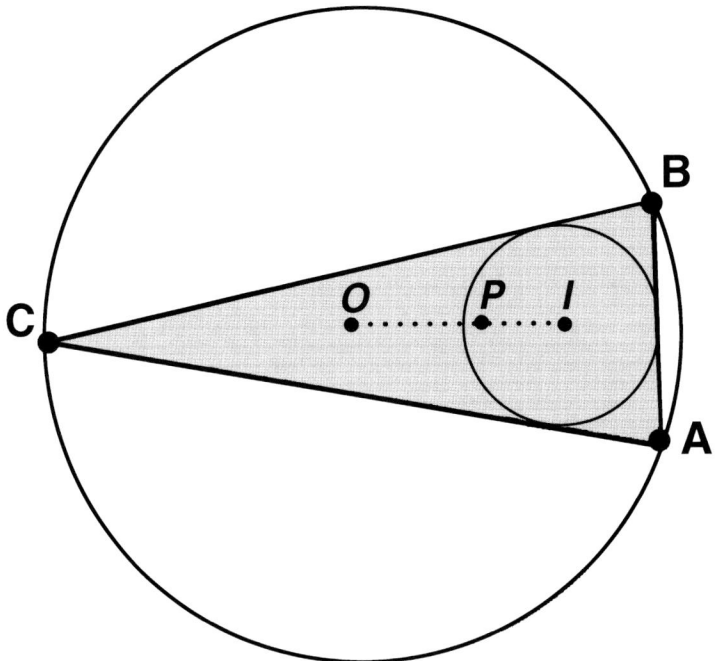

Figure 15. Generation of a new node P inside a selected triangle ABC, with I and O the centres of the inscribing and circumscribing circle of ABC.

– Generate a new node *P* inside the selected triangle using the following formula:

$$P = (1 - p) * \text{center of inscribing circle} \qquad [4]$$
$$+ p * \text{center of circumscribing circle}$$

with:

$$p = 2 * \text{radius of inscribing circle/} \qquad [5]$$
$$\text{radius of circumscribing circle}$$

The value of *p* reaches its maximum of one if the triangle is equilateral, and is smaller if the triangle is more 'stretched' (see Figure 15).

– Calculate the value of $d(x, y)$ in the new node *P*. Discard *P* if the distance of *P* to any vertex of the triangle is too small.

- Determine which triangles of the triangulation can be removed if *P* is inserted. If the insertion polygon contains only one triangle, *P* is discarded, because insertion of such a node would cause triangles with obtuse angles to be formed.
- After accepting node *P*, update the Delaunay triangulation.

The insertion of nodes to the interior of the object is stopped when all triangles satisfy the mesh density function.

The 2D algorithm can be extended to 3D as follows:[18]

- Generate nodes on all edges of the object, complying with the (3D) density function.
- Insert nodes into all faces of the object using the 2D mesh generator.
- Construct a Delaunay tetrahedrization for the nodes on the boundary of the object.
- Insert nodes to the interior of the object by using a 3D version of the above method.

3.4.3. Conclusions

The advancing front method[15–17] is suited to the generation of meshes in both 2D and 3D, although the method has some disadvantages. The generation of interior nodes and the selection of an interior node have no firm theoretical basis, and no optimality criteria are satisfied. This may lead to the creation of badly-shaped elements when the advancing front consists of few edges or triangles, and 'collapses' onto itself. In that case, the distances between the nodes have become so small that only nodes on the front itself can be used to create a new element. The selected node may then result in a badly-shaped element.

The Delaunay triangulation method of Frey[18] has as main advantages that it does have a theoretical basis, and that, at least in 2D, it usually gives good meshes.

3.5. MESH GENERATION BASED ON OBJECT OR OBJECT SPACE DECOMPOSITION

These methods start by decomposing either the object or the object space into simpler pieces, which can then be meshed more easily. A mesh is then generated for these simpler pieces.

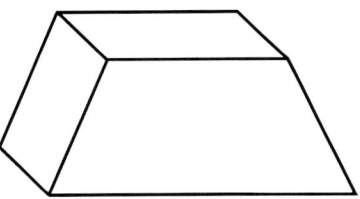 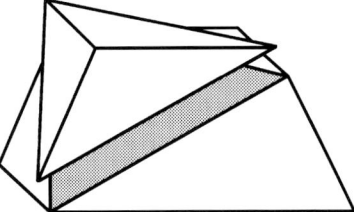

Figure 16. Operation τ_1 applied to a polyhedron.

3.5.1. Mesh Generation by Topology Decomposition

An object is first decomposed into a number of crude tetrahedral elements, using only the vertices of the object as nodes for the mesh. These crude elements are then refined by the introduction of new nodes.

Woo and Thomasma[19] describe an algorithm for the tetrahedrization of a object by topology decomposition, where three operators are used to decompose the object. When creating the tetrahedra, the shape of the elements is not considered, and thus badly-shaped elements may be formed.

The algorithm first generates elements using the operators τ_1 and τ_2. Operator τ_1 cuts a tetrahedral elementfrom the object if it contains a convex vertex where three edges meet (see Figure 16). If operator τ_1 cannot be used, operator τ_2 is used to cut a tetrahedral element by making two incisions into a convex edge of the object (see Figure 17). Since a polyhedron always contains a convex edge, operator τ_2 can always be used.

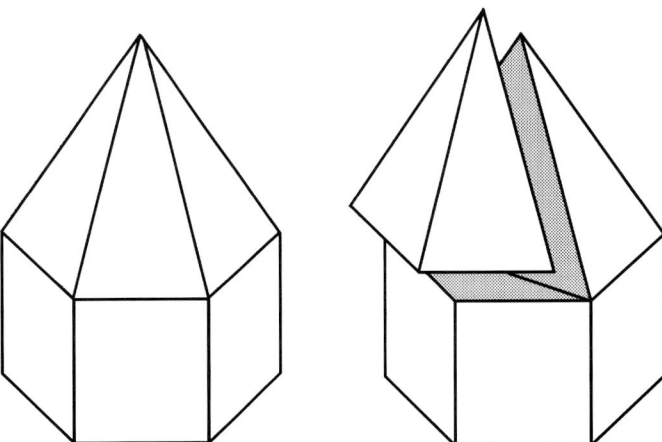

Figure 17. Operation τ_2 applied to a polyhedron.

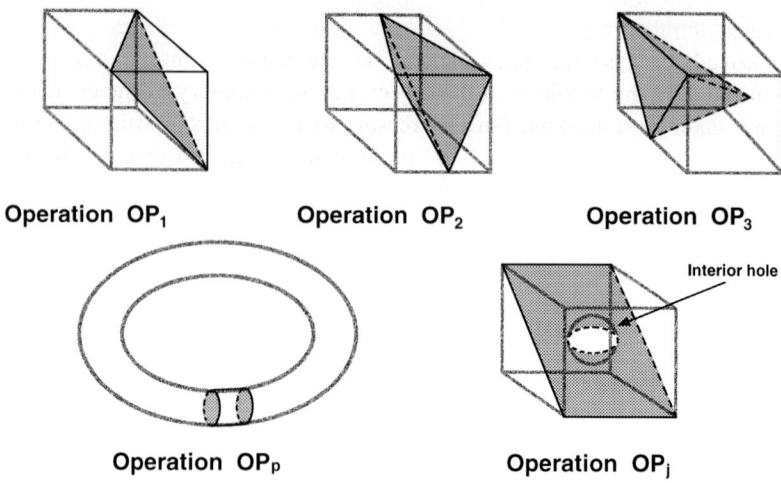

Figure 18. Tetrahedrization of a 3D object using the operations of Wördenweber.

Solid objects without holes in their interior can be tetrahedrized using only τ_1 and τ_2, but holes in the interior have to be eliminated using a third operator, τ_3, which 'opens' the object by cutting a number of tetrahedrons from the object. After applying operator τ_3, τ_1 and τ_2 can be used to decompose the remainder of the object. When the object has been decomposed into crude tetrahedral elements, the second step of the algorithm subdivides these tetrahedral elements.

Wördenweber[20,21] tetrahedrizes an object by first triangulating the faces of the object, and then using a number of operators to cut tetrahedrons from the object. The following operators are used (see Figure 18):

–   Operator OP_j, for elimination of holes in the interior by opening the object.
–   Operator OP_p, for opening a toroid.
–   Operators OP_i, $i = 1, 2, 3$ for cutting a tetrahedron from the object. This is done along i triangles, thereby introducing i new triangular faces in the boundary representation. The use of operator OP_i is therefore preferred to that of OP_{i+1}.

The result of the operators on the object is again a set of crude tetrahedra. These tetrahedrons are refined by subdivision, and their shape is improved by repositioning the nodes.

The predominant disadvantage of topology decomposition is that the created tetrahedrons are often badly shaped, and cannot be sufficiently improved, because the generation of the elements is almost completely controlled by the topology of the object, i.e. the geometry of the elements is not taken into account. For this reason, topology decomposition can in general not be recommended for mesh generation. However, a special version of topology decomposition, called element extraction, has been successfully applied for meshing boundary octants of an octree.[22]

3.5.2. Mesh Generation Using Spatial Decomposition

Methods for mesh generation using spatial decomposition, perform a regular subdivision of the space surrounding an object. The decomposition results in a number of regions in space that can be classified as inside or outside the object, or intersected by the boundary of the object. These regions can then be used to localize the meshing process. The principal aim of the subdivision process is to replace meshing of a complex object by meshing of regions in which the object can be considered 'simple'. Five methods are discussed: mesh generation using a quadtree approach, an octree approach, an octree and Delaunay tetrahedrization, an octree and element removal, and octrees with tetrahedral octants.

3.5.3. Mesh Generation Using a Quadtree Approach

The quadtree is a hierarchical data structure that is based on recursive decomposition of 2D space.[23] The approximating quadtree representation of a 2D object is derived by recursive subdivision of a square surrounding the object into cells or quadrants. Each subdivision creates four new quadrants from an existing one. The subdivision of a quadrant continues until either it can be decided that the quadrant is inside or outside the object, or the maximum subdivision level has been reached. Thus, the quadrants are usually of different sizes. An example of a quadtree can be seen in Figure 19.

So a quadtree can be used to subdivide an object into quadrants, in which the meshing of an object can be localized. In[24] cut quadrants are used to better approximate the boundary of an object. Cut quadrants are quadrants cut by a single line segment, connecting the quarter, half, or end points of any two sides. They are thus restricted to a limited number of configurations (see Figure 20 for a number of examples). The configuration to use in a particular case is determined by the computed intersections between quadrant edges and the object. This modified quadtree therefore still cannot be expected to exactly represent the boundary of the object, and the same applies to a mesh derived from the modified quadtree.

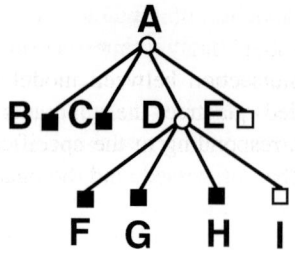

Figure 19. Example of a quadtree.

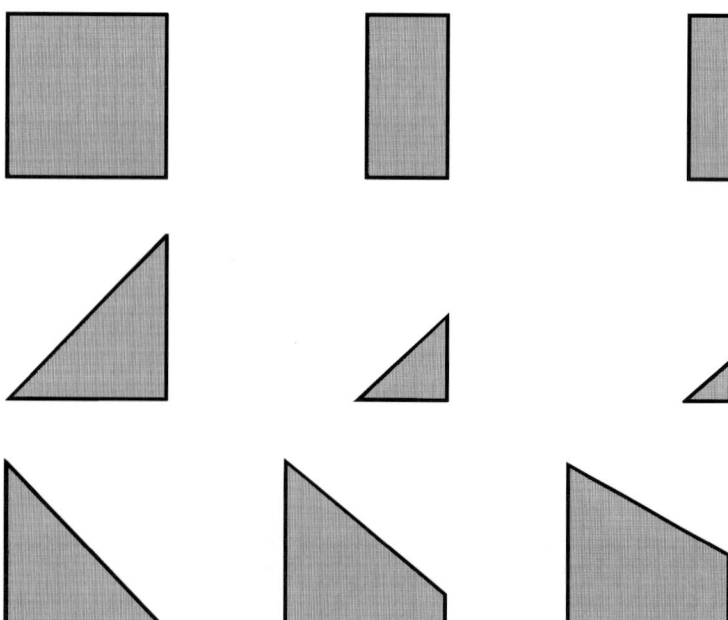

Figure 20. A number of cut quadrants.

An improvement on the above method is given in,[25] which allows explicit storage of vertex locations and boundary intersection nodes obtained from the computation of the intersection between model and quadrant edges. When deriving the modified quadtree, the quadrants are first subdivided until a level is reached corresponding to the specified mesh density. The boundary of the model is then processed, and the quadrants are filled with the geometric information from the model, including pointers to the intersecting edges of the model. The modified quadtree thus obtained contains the complete geometric information about the object. A quadrant may contain information about more than a single vertex or edge, and therefore geometry of arbitrary complexity can be handled.

The modified quadtree representation may contain small segments resulting from object boundaries intersecting a quadrant close to a corner, or object vertices lying close to a quadrant side. Meshing such a quadrant would lead to very small or distorted elements. Therefore, a technique was developed to pull object vertices close to a quadrant side towards that side, and boundary intersection nodes towards quadrant corners. It should be noted that this technique only changes the modified quadtree, but leaves the original geometric description of the object unchanged.

Finally, the quadrants are meshed on an individual basis. Interior quadrants may be meshed using templates. A mesh template is a pregenerated mesh that is adapted to the quadrant being meshed. However, boundary quadrants may be of arbitrary complexity, and may therefore need an algorithm for meshing an arbitrary 2D object. This is obviously the main drawback of the method: the geometry contained within a quadrant is not guaranteed to be less complex than the geometry of the object itself.

More recently, spatial subdivision techniques have also been used to compute nonobtuse and no-small-angle triangulations of 2D node sets or 2D objects. An overview of algorithms to compute such optimal triangles is given in.[26] Bern et al.[27] describe quadtree-based techniques that produce a triangulation of a 2D node set with neither obtuse nor small angles, and techniques that produce a triangulation of a polygon without small angles. Baker et al.[28] and Melissaratos and Souvaine[29] describe algorithms that produce a triangulation of a polygon with neither obtuse nor small angles; these algorithms are based on a combination of spatial subdivision and geometry decomposition. Geometry decomposition is a technique that cuts elements from an object that can be simply triangulated by well-shaped triangles, and is controlled by the topology and the geometry of the object. All algorithms that guarantee to produce a nonobtuse or no-small-angle triangulation introduce extra nodes to the node set or polygon.

3.5.4. Mesh Generation Using an Octree Approach

The generalization of the quadtree to 3D is the octree. To obtain an approximating octree representation with cubical cells, a cube surrounding the object is recursively decomposed into octants. Each octant subdivision creates eight smaller octants.

To generate a mesh for 3D objects, the modified octree can be used, just as the modified quadtree approach is used for 2D objects. After building the octree, the octants in the modified octree are meshed. The modified octree presented in[30] is a direct generalization of the modified quadtree. Instead of cut quadrants, cut octants are used, and again only a limited number of configurations are possible. The modified octree representation suffers from the same deficiencies as the modified quadtree approach.

3.5.5. Mesh Generation Using an Octree and Delaunay Tetrahedrization

A different approach to the use of octrees is presented in.[31,32] Here, the octree is used to localize the Delaunay tetrahedrization of nodes on or within an octant to that specific octant. First, the geometry of the object is intersected with the octants, and the octree is built. Then a Delaunay tetrahedrization is built within each octant. Since a Delaunay tetrahedrization yields the tetrahedrization of the convex hull of the node set, the resulting tetrahedrization may not be compatible with the part of the object inside the octant. Incompatibilities (see Figure 21 for an example of a triangulation

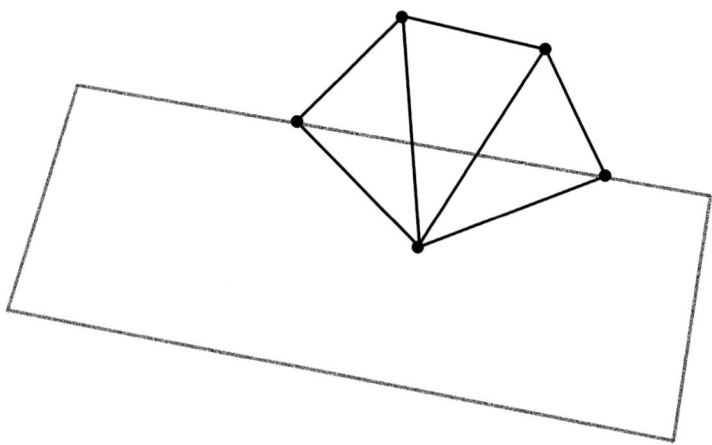

Figure 21.   Triangulation incompatible with a face.

incompatible with a face) are detected by classification of the mesh entities (vertices, edges and faces) with respect to the geometric model. Thus, a complete geometric representation should be made available for each octant, so that local classification and compatibility checking can be performed. Checking between octants is necessary to ensure compatibility of the octant face triangulations.[32] A final disadvantage is that the resulting mesh will not satisfy the Delaunay properties for the whole object, since the octants are meshed individually.

In[33] an improved version of the above method is discussed. It consists of the following steps:

–   The octree is built by intersecting the edges and the faces of the octants with the geometry of the model to produce a set of intersection nodes. The vertices of the model are inserted into the octree as well. The vertices of the octree cells are classified with respect to the model, and each octant is classified as inside, outside or on the boundary.
–   For each octant inside or on the boundary, a Delaunay tetrahedrization is generated involving only the vertices of the octant. Together these tetrahedrizations form a Delaunay tetrahedrization covering the entire object.
–   The intersection nodes and vertices of the model are now inserted into the octree tetrahedrization.
–   After completion of the Delaunay tetrahedrization, the mesh entities are classified with respect to the model geometry, and topological incompatibilities are resolved.

The advantage of this method is that it results in a Delaunay tetrahedrization of the object, and that it is not necessary to build a complete geometric representation for each octant. Important disadvantages that remain are the complex compatibility checking, and the need to resolve incompatibilities by the introduction of new nodes, or by octree refinement.[31,33]

3.5.6.   Mesh Generation Using an Octree and Element Removal

In[32,33] a 3D mesh is built from the information in the octants without further regarding the topology and geometry of the model. Instead, the tetrahedrization is adapted afterwards. In[22] a different approach is taken. During the creation of the octree, in each octant a polygonal approximation is built that fully represents the topology and the geometry of the model in that octant. The octree is thus a complete representation of the geometric object, partitioned into a number of octants, with each octant containing a topologically correct approximation of the part of the object intersecting that octant.

The FE mesh is derived from the geometric model in two steps. First, the octree is built. Second, the mesh is derived from the octree. Building the octree involves insertion of the vertices, edges and faces into the octree. The intersection computations between boundary elements of the geometric model and edges and faces of the octants are performed by the geometric modeller to ensure compatible results. From the inserted vertices and computed intersections, a geometric representation is built in each boundary octant. The topological elements in this representation have to be checked for topological compatibility and geometric similarity against the model. If an inconsistency arises during the geometry building of an octant, geometric operations are performed, or an octant is subdivided. The meshing process itself uses an element removal technique, a special version of topology decomposition, to mesh the boundary octants in the tree.

3.5.7. Mesh Generation Using Octrees with Tetrahedral Octants

In[34] an octree approach with tetrahedral instead of cubical octants is used. A tetrahedral octant is divided into eight smaller tetrahedra, such that the eight tetrahedrons are Delaunay tetrahedra. The solid object to be tetrahedrized is supposed to have its faces triangulated. With the triangulation of the faces, two sets of spheres are defined. One set consists of the spheres of minimum radius (minspheres) passing through the vertices of each triangle. The second set consists of spheres centred on each vertex of the boundary triangulation, the radius of such a sphere being given by the mean of the radii of the mini spheres corresponding to the triangles adjacent to that vertex. Together, these two sets of spheres define an irregular offset of the boundary. No interior nodes between the boundary and the offset are allowed. When building the octree representation, octants that intersect the spheres are classified as boundary octants, and subdivided until a maximum subdivision level has been reached. This maximum subdivision level can be different for different octants, and is partially dependent on the size of the triangles on the faces. If an octant intersects a sphere when the maximum subdivision is reached, it is classified as out. After the octree has been built, the vertices of the boundary triangulation and the vertices of the octants, which are Delaunay tetrahedra, are used to generate the Delaunay tetrahedrization.

A disadvantage of the method is that the local boundary triangulation of the model may not be reproduced by the tetrahedrization, unless it is a local boundary Delaunay triangulation, i.e. the minsphere of each triangle contains no nodes in its interior. This implies that the obtained Delaunay tetrahedrization will not be guaranteed to represent the boundary of the object correctly for all boundary triangulations.

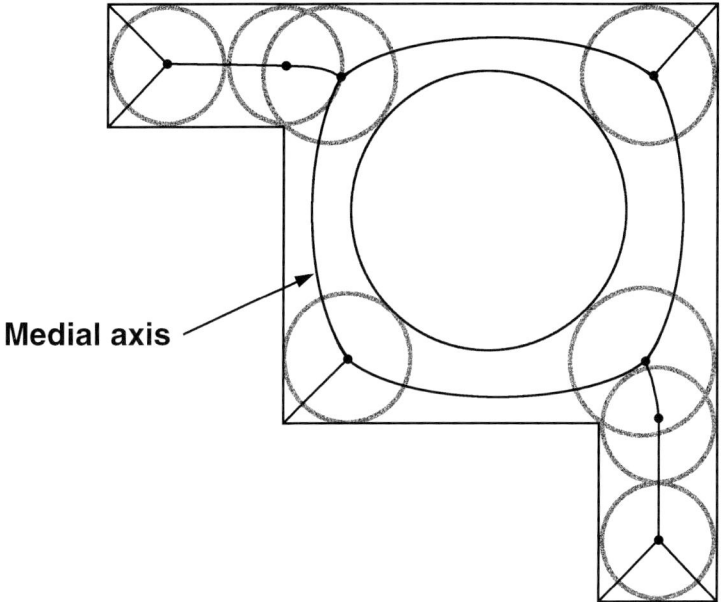

Figure 22. Medial axis of a 2D object.

3.5.8. Mesh Generation by Medial Axis Subdivision

Medial axis subdivision[35] is a mesh generation method that uses information derived from the so-called medial axis of a 2D object. The medial axis is defined as the locus of the center of a circle of maximum size as it moves along the inside of the object (see Figure 22). The information derived from the medial axis is used to subdivide the object into subregions that are simple enough to mesh directly.

The medial axis can be obtained from the Voronoi diagram of an object, removing all edges of the diagram that originate from concave vertices of the object. The Voronoi diagram of the edges is the set of Voronoi regions associated with these edges. A Voronoi region associated with an edge is the set of points that are closer to that edge than to any other part of the boundary.

An approximation of the Voronoi diagram can be obtained from the Delaunay triangulation of a set of nodes distributed along the boundary of the object. Only the vertices of the Voronoi diagram contain features of interest for the generation of the mesh. So only the Delaunay triangles

Figure 23. Delaunay triangle classification of a 2D object.

corresponding to those points are generated. These Delaunay triangles can be classified into one of the following five types (see Figure 23):

– The junction type J.
– The intermediate type I.
– The corner type C.
– The normal type N.
– The topologically redundant type T.

The topologically redundant type T has all its vertices on a curved edge of a boundary, and is not shown in Figure 23.

After calculating the medial axis of an object, the object can be split into simple subregions by using a series of operations to subdivide its medial axis. The subregions are meshed with a suitable mesh type and pattern. The resulting meshes are well structured and have a high quality.

This algorithm can be extended to 3D by using the concept of medial surface subdivision. The medial surface of a 3D object is the locus of the center of a sphere of maximum size, moving around inside the 3D object. However, in[35] it is mentioned that this concept is still under investigation.

3.5.9. Conclusions

The predominant disadvantage of topology decomposition[19–21] is that the tetrahedrons created are often badly shaped and cannot be sufficiently improved. This is caused by the fact that the generation of the elements is almost completely controlled by the topology of the object, i.e. the geometry of the elements is not taken into account. For this reason, topology decomposition can in general not be recommended for mesh generation. However, a special version of topology decomposition, element extraction, has been successfully used for meshing boundary octants of an octree.[22]

The main advantage of the spatial subdivision approaches[23–34] to mesh generation is that they usually generate cells that are more easily meshed than the original object. A disadvantage of the quadtree and octree techniques is that the obtained subdivision of the object is not generated directly from the geometry of the object itself, but, instead, it results from a spatial subdivision. This implies, among other things, that model vertices can be arbitrary close to the boundaries of a cell, and that direct meshing of the cells can thus result in many small and deformed elements. Therefore, the obtained representation has to be adapted to prevent the creation of such elements. The second, and most important, disadvantage of the spatial subdivision techniques is that the geometry of an object within a cell is not guaranteed to be less complex than that of the object itself; subdivision of a cell does not necessarily yield cells that can be meshed more easily.

The resulting meshes of the medial axis subdivision method[35] are well structured and have a high quality. However, the algorithm has only been implemented for 2D. It might be extendible to 3D, by using the concept of medial surface subdivision, but this is still under investigation.

So, if 2D objects have to be meshed by one of the above methods, use of the medial axis subdivision method should be favoured. However, if the objects to be meshed are 3D, a choice has to be made between topology decomposition and spatial decomposition. In this case spatial decomposition seems to be the best choice.

3.6. MESH GENERATION FROM CSG MODELS

In the previous sections, it was assumed that objects are represented by a boundary representation, i.e. that information about the faces, edges and vertices, including the adjacencies between these entities, was stored in the geometric model of an object to be meshed. Mostly it was also assumed that curved facesof objects had been approximated by planar faces, because this makes the algorithms considerably simpler. An alternative is to represent objects with Constructive Solid Geometry (CSG), and to allow curved faces in such a model.

CSG is based on the use of, usually rather simple, primitive solids, such as cubes, cylinders and spheres. Instances of these primitives can be created, and a combination of translations, rotations and scalings can be applied to them. These primitives are then combined with the set operations union, difference and intersection, to form a more complex composite object.

A CSG model is represented by a binary tree, the CSG tree. At a leaf or primitive node, information about the primitive is stored, which consists of its type and the geometric transformations applied to the primitive to get it in the correct position, orientation and size relative to the other primitives composing the object. At an internal or operation node, the set operation applied to the objects in the left and right subtrees is specified. So there is no explicit information about the faces, edges and vertices of the resulting object, only information about which primitives are combined, and how these are combined.

A CSG model can be interrogated in terms of the classification of 3D points as being inside, outside, or on the boundary of the composite object. A point can first be classified with respect to the primitives in the leaf nodes. These classifications can then be combined at the operation nodes while traversing the tree, finally yielding the classification with respect to the composite object.

A CSG model can be converted into a boundary representation by a process called boundary evaluation, in which the faces, edges and vertices of the model are computed from the primitives and set operations in the CSG model. This is, however, a complicated, time consuming and error prone process. Therefore in several applications attempts have been made to avoid boundary evaluation, and work directly on the CSG model. Here an overview is given of attempts to generate a mesh directly from a CSG model.

In[36] a description of a method is given to derive a 2D mesh from a CSG model with 2D primitives. The algorithm distinguishes nodes on different topological entities, and separates the generation of the nodes from that of the elements. It consists of the following three steps:

- Generate a good distribution of nodes on the boundary of and inside the primitives (see Figure 24 for an example).
- At each operation node of the CSG tree, combine the node sets of the left and right subtree according to the specific operation (see Figure 25 and 26 for an example).
- Create the elements after the node set for the root node of the CSG tree has been determined.

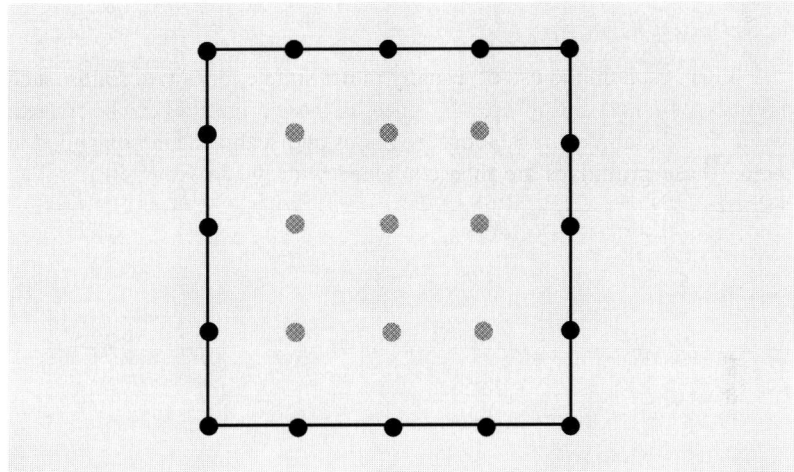

Figure 24. Distribution of nodes on the boundary of (black dots) and inside (grey dots) a primitive.

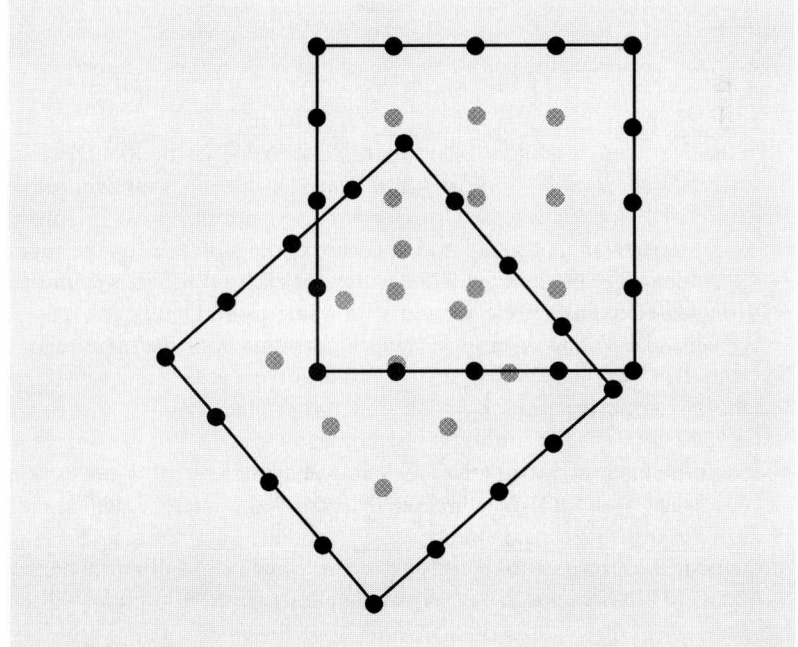

Figure 25. Two node sets to be combined under the union operator.

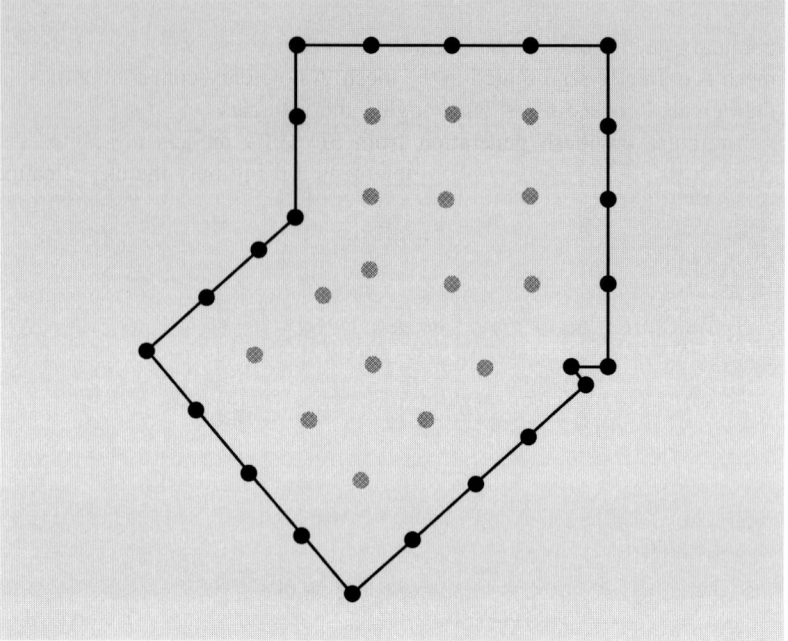

Figure 26.   The resulting node set after combination under the union operator.

In the first step, a number of nodes are distributed on the boundary of each primitive, called *b*-nodes, and in the interior of each primitive, called *i*-nodes. The *b*-nodes are further subdivided into *v*-nodes, corresponding to the vertices of the model, and *e*-nodes, corresponding to nodes on the edges of the model. The distribution of the nodes for each primitive is such that well-shaped elements can be formed within that primitive.

The second step first determines at each operation node the intersections between the object in the left tree (LT) and the object in the right tree (RT). For these intersections, new *v*-nodes are created. Then the node sets of LT and RT are combined according to the operation specified at the operation node. Combining nodes involves deletion and merging of *b*-nodes and *i*-nodes from LT and RT. Whether a node is deleted or merged with a node of the other tree, depends on the types of both nodes, and the classification of the node with respect to the other tree. An *i*-node close to a *b*-node, for example, will be deleted, whereas two *i*-nodes that are too close will be merged.

In fact, a partial boundary evaluation is performed at the second step, so that for each *b*-node its neighbouring *b*-nodes are known. This implies that during the third step, the creation of the elements, it can be determined which boundary nodes should be connected to guarantee that the boundary of the mesh is correctly represented in the mesh. A boundary-conforming mesh is then created consisting of quadrangles and triangles.

Literature on mesh generation from 3D CSG models is scarce. The algorithm described in[37] can be implemented using only the classification of potential nodes on and in each primitive against the CSG tree, but no nodes for the vertices or on the edges of the solid are generated, and the resulting tetrahedrization therefore cannot be expected to represent the boundary of the solid correctly. Other descriptions of mesh generators based on 3D CSG modellers assume the derivation of a boundary representation from the CSG model before the object is meshed.[14,38]

The 2D algorithm described in[36] is a node-based approach; the node set for a CSG tree is obtained from the node sets for the subtrees, and the computed intersections between edges. However, the mesh generation phase, following the node generation phase, requires the adjacencies between boundary nodes to determine the boundary of the mesh. These adjacencies are determined during the node generation phase. A generalization of this phase to 3D would require the computation of the vertices and edges of the 3D solid, and the adjacency relations between vertices, edges and faces. Thus, an almost complete boundary representation is required. It is claimed that the mesh generator described in[22] does not require the derivation of a complete boundary representation, but only some basic topological entities and relations. However, it is not clear from the article which basic topological entities and relations are required by the algorithm.

In[39,40] it is shown that the derivation of a mesh that is a topologically correct discretization of a CSG model actually requires a boundary representation of the model. In the approach that is presented, derivation of a mesh from a CSG model has therefore been divided into two steps. The first step is boundary evaluation of the CSG model, the second step is generating a tetrahedral mesh from this boundary representation.

An approach to boundary evaluation has been chosen that reduces all computations on edges and faces to 2D. The primitives in a CSG model have a dual representation; their faces have both a representation by rational Bézier patches and a representation by algebraic surfaces. The faces in a boundary representation are represented by trimmed Bézier patches, and vertices and edges have a local representation with respect to each neighbouring face.

An essential part of every boundary evaluation procedure, is a method to derive the intersection curve between intersecting faces of two primitives. It was decided to represent the intersection curve as an algebraic curve in the parameter spaces of the intersecting faces. In this way, the intersection curve can in general be derived with high accuracy.

The meshing of the boundary representation has been divided into two steps. The first involves triangulation of the faces, with the faces defined as trimmed patches. The second consists of tetrahedrization of the solid, starting from the triangulation of the faces.

An algorithm for triangulating bounded planar domains has been implemented, which is used to triangulate the faces in the boundary representation. This algorithm can be used for arbitrary non-convex domains. A constrained Delaunay triangulation is first generated that is a topologically correct representation of the domain, i.e. all edges of the domain have a corresponding representation in the triangulation. This triangulation is then optimized by adjusting its topology and the positions of its nodes. Tetrahedrization of the object, starting from the triangulation of the boundary, has not been implemented.

The main conclusion of this research is that the representation of faces as trimmed patches is very suitable, both to boundary evaluation of exact models and to mesh generation from the resulting boundary representation, and thus to mesh generation from exact CSG models. The mesh generation algorithm ensures that a valid triangulation of the boundary representation is generated, and the triangles obtained are usually well shaped. The reliability of the boundary evaluation procedure, however, needs to be further improved.

3.7. MESH GENERATION FROM FEATURE MODELS

Feature modelling is a relatively new development in CAD. Whereas in solid modelling only information about the geometry of objects is considered, in feature modelling also functional information is stored in a model. The functional information may be the function for the user of some part of the object, or the way some part of the object is manufactured,but also information relevant for analysis purposes, as will be shown here. A survey on feature modelling can be found in.[41]

A feature is a part of an object about which geometric and functional information is stored. Often, the functional information of a part is more important than the exact shape of it. However, form still plays a predominant role in many features. Slots, protrusions, pockets and holes are good examples of features.

A distinction can be made between elementary and compound features. Elementary features are simple features that cannot be decomposed into still simpler features. An example of an elementary feature is a hole. Compound features are more complex features composed of several elementary features. An example of a compound feature is a stepped hole, consisting of two concentric holes.

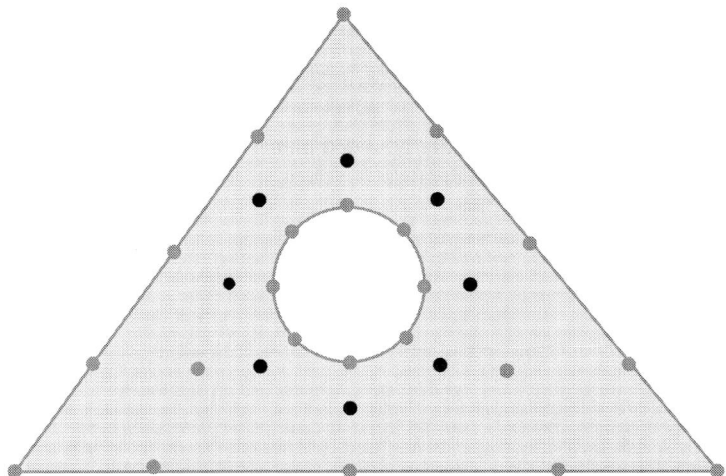

Figure 27. Nodes inserted into the region around a hole feature (black dots).

Features can be exploited in the mesh generation process, because they provide a means of linking information relevant for analysis to the object. Boundary conditions can be linked directly to features of the object, and features can make explicit important regions of the object, and this can be used to steer the mesh generation.

In[42] an approach to finite element mesh generation using a feature-based model is described. First nodes are generated for each feature. Each feature has his own mesh density, so the mesh density can be controlled locally. If features intersect, nodes are generated on the intersection using the finest of the mesh densities. Also, nodes are generated around the features. These are used to reconcile the feature's local mesh density with the global mesh density, as shown in Figure 27. Not all generated nodes are used. Nodes outside the object or too close to the boundary of other features are removed (see Figure 28).

When all nodes have been generated, they are used to create a Delaunay triangulation. The created mesh is usually consistent with the original object. Any inconsistencies are detected in the verification stage, in which holes in the mesh that are not present in the object, and boundary intersections of the mesh with the object, are detected. The holes can be corrected by inserting extra nodes into the triangulation, and the intersections can be corrected by removing elements from the mesh as described in.[33] After all inconsistencies have been removed, the mesh is smoothed to improve the mesh quality. This is done without moving nodes from their associated

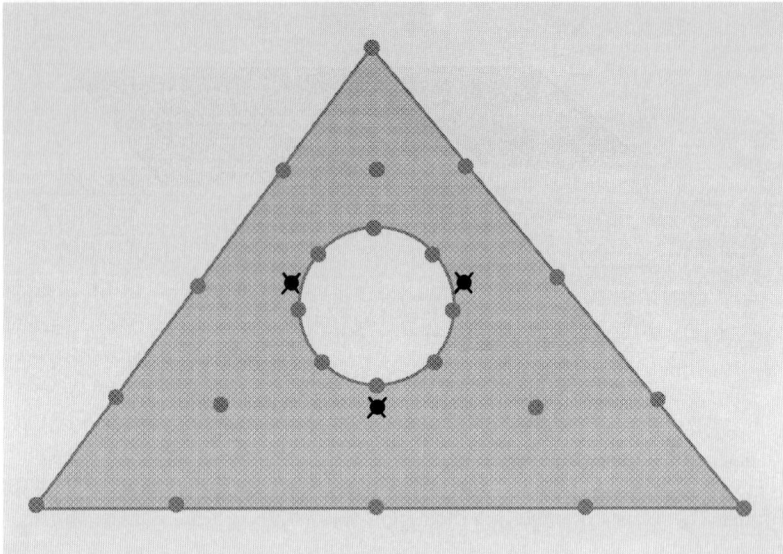

Figure 28.   Nodes of a hole feature that are removed because they are too close to the boundary of the triangular feature.

boundary element. So nodes generated on an edge, are only moved along that edge during the smoothing process, and nodes generated on a face, are only moved along that face. The associated boundary element of the nodes was stored during the generation of the nodes.

The advantages of this approach are that the mesh density can be controlled in a natural way, and that the algorithm is simple.

In[43] a feature-based mesh generation algorithm is described for functional surfaces. Functional surfaces contain compound features, mostly consisting of pockets, channels and ribs. An example of such a surface is an automobile inner panel.

Elementary features are defined by a secondary surface S_2, for example the floor of a pocket, that must be linked to a primary surface S_1. On S_1 the curve C_1' is defined, and on S_2 the curve C_2' (see Figure 29). A smooth transition between the two curves links the surfaces. The projections of C_1' and C_2' onto the xy-plane, C_1 and C_2 respectively, divide the plane into three regions R_1, R_2 and R_T. The region R_1 is outside C_1, the region R_2 is inside C_2, and R_T is the transition region between C_1 and C_2.

Assume that S_1 and S_2 can be described by the functions $f_1(x, y)$ and $f_2(x, y)$ respectively, and a transition function g is defined as follows:

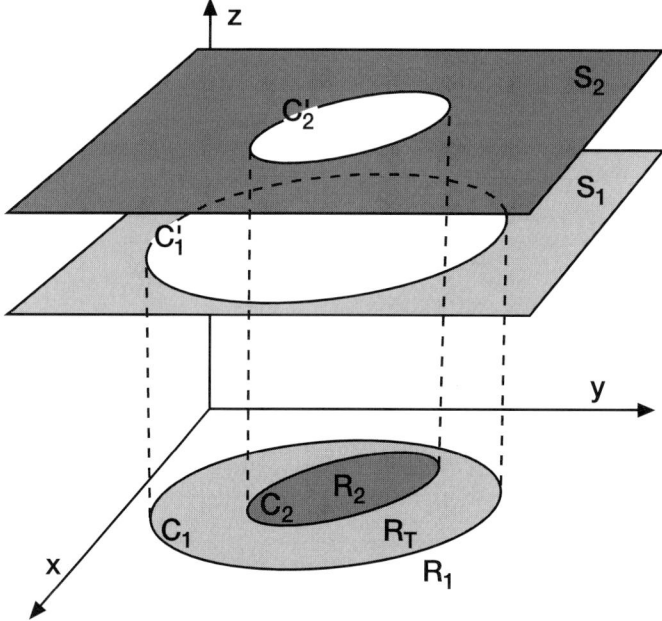

Figure 29. Elementary feature defined by a primary and secondary surface.

$$g(x, y) = \begin{cases} 0 & \text{if } (x, y) \in R_1 \\ \in (0,1) & \text{if } (x, y) \in R_T \\ 1 & \text{if } (x, y) \in R_2 \end{cases} \qquad (6)$$

then the feature surface can be described by

$$h(x, y) = (1 - g(x, y)) f_1(x, y) + g(x, y) f_2(x, y) \qquad (7)$$

Compound features are defined as a series of N features whose tops or bottoms are defined by surfaces S_i described by $f_i(x, y)$, i = 1...N, on a base surface S_0. For each feature i, there is a transition function $g_i(x, y)$ corresponding to the shape of the ith feature. The total surface $h_N(x, y)$ can now be defined by recursive use of

$$h_i(x, y) = (1 - g_i(x, y)) h_{i-1}(x, y) + g_i(x, y) f_i(x, y), 1 \leq i \leq N \qquad (8)$$

where $h_0(x, y) = f_0(x, y)$.

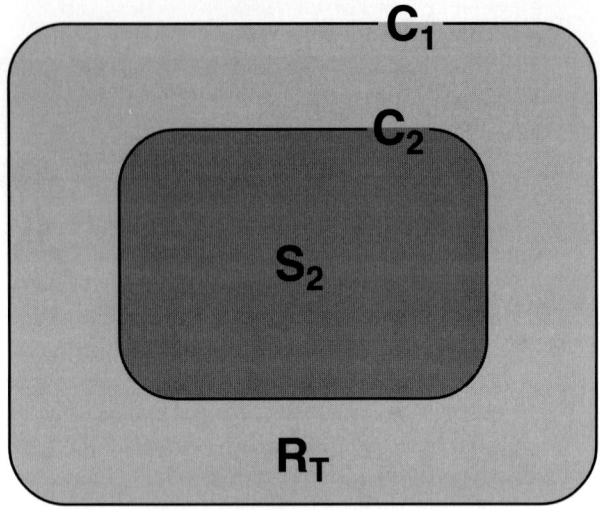

Figure 30.   Surfaces of an elementary feature to be meshed.

In[43] the features on the 3D surface are defined by the user in 2D. Mesh generation takes place by mapping a planar triangulation onto the surface, by projecting all the nodes of the mesh onto the 3D surface. The mesh generator generates planar meshes for each feature individually.

For an elementary feature, the transition region R_T and the secondary surface S_2 are disjoint, and can be meshed independently (see Figure 30). When meshing S_2, the boundary C_2 is seen as the outer boundary of S_2. When meshing R_T, it is seen as the boundary of a hole. The regions are meshed as follows:

–   Nodes are generated along the boundaries C_1 and C_2.
–   Additional uniformly spaced curves are generated in the transition region between C_1 and C_2 (see Figure 31).
–   Nodes are generated along these new curves (see Figure 32).
–   Once all these nodes have been generated, R_T and S_2 can be triangulated independently, because the nodes on the boundary C_2 are the same for both regions.

Generalization of elementary feature meshing to compound feature meshing is easy. Compound features can be classified into three categories:

1.   Isolated features, which lie on the secondary surface of another feature, but are isolated from all other features.

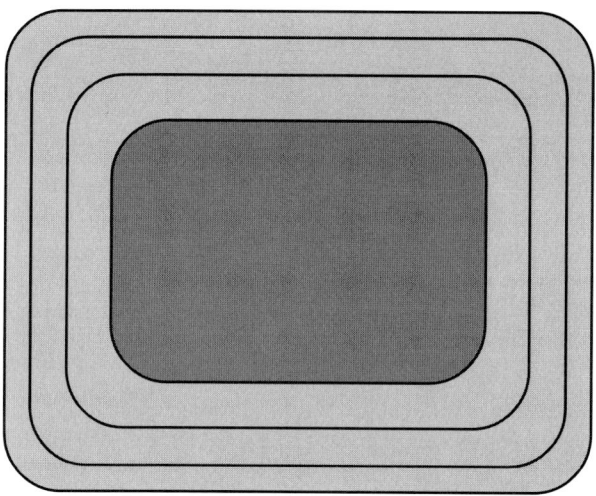

Figure 31. Additional uniformly spaced curves in the transition region of the feature.

2. Separable features, which contain other features on their secondary surface.
3. Truncated features, which partially overlap other features.

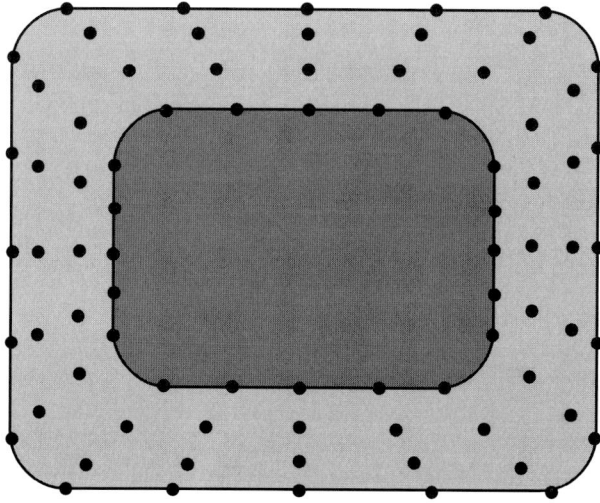

Figure 32. Nodes generated along the curves.

The triangulation of these features consists of the following three steps:

1. Each isolated feature is triangulated in the same way as an elementary feature.
2. Each separable feature is also triangulated in the same way as an elementary feature. However, the boundaries of the features contained inside this feature are treated as holes and any truncated features intersecting this feature as excisions.
3. The remainder of the surface, including the truncated features, is triangulated by treating the boundaries of the isolated and separable features contained inside the remainder of the surface as holes.

So this mesh generator again generates meshes for each feature individually, which has as advantage that each mesh is determined according to local requirements. Together the meshes form one consistent Delaunay triangulation of the entire surface.

In both approaches discussed in this section, it has been shown that features can be useful to steer the mesh generation process, and their use is therefore an important new development in mesh generation.

3.8. MESH IMPROVEMENT TECHNIQUES

Several techniques have been developed that aim to improve the shape of the elements of an existing mesh. These methods are usually iterative. The following types of methods can be distinguished:

- Methods that change the geometry of the mesh in each iteration.
- Methods that change the topology of the mesh in each iteration.
- Methods that change both geometry and topology in each iteration.

The most important method of the first type is the Laplacian method, which aims to improve the shape of the elements in a triangulation by repositioning the interior nodes. A new position P_{new} for an interior node P_i is computed using the following formula:

$$P_{new} = \frac{1}{N_i} \sum_{j=1}^{N_i} P_j \qquad (9)$$

with P_j being the adjacent nodes to P_i, and N_i the number of adjacent nodes.

A straightforward application of this formula to compute the smoothed positions of interior nodes, may result in interior nodes outside or on the boundary of the triangulation, and thus in intersecting and partially overlapping triangles. This is prevented in the following way:

- After the computation of P_{new}, it is tested whether P_{new} falls in the interior of the kernel of UP_i, UP_i being the polygon defined as the union of all triangles adjacent to P_i. The kernel of a polygon is defined as all points that are visible from all edges of the polygon.
- If P_{new} is not in the interior of the kernel of UP_i, a new P_{new} is computed as a weighed average of P and P_{new}. This step may have to be repeated several times.

By computing P_i this way, it will always remain in the interior of the triangulation, and triangles will not overlap. The method can be used for both 2D and 3D meshes. The basic method can be adapted for repositioning boundary nodes, in which case a new position has to be constrained to the boundary.

The topology of a mesh can be changed by adjusting the connection structure of the mesh. In 2D, the topology of a mesh can be changed by edge swapping. An edge shared by two triangles is replaced by an edge connecting the other two nodes of the triangles. Similarly, in 3D, face swapping can be used.

A generalization of the edge swapping algorithm, called edge insertion, has been introduced in[44] to minimize the maximum angle of a triangulation. This technique repeatedly attempts to eliminate the largest angle PQR from a triangulation by insertion of an edge E, connecting Q with another node of the triangulation. The edges of the existing triangulation crossing the new edge E are subsequently removed, which creates two simple polygons incident with E. These two polygons are then triangulated in such a way that their maximum angle is minimized.

In[45] a technique called mesh relaxation is described to iteratively improve the regularity of a triangulation by edge swapping. The regularity of a triangulation is defined by the number of edges connected to each triangulation node, which is called the degree of the node. A n-node is a node of degree n. The regularity of a triangulation can be improved by reducing the degree variations of the nodes. In a regular triangulation consisting of equilateral triangles, all interior nodes have degree six. On the basis of an ideal degree of six for each interior node, a criterion can be developed that determines for each interior edge of the triangulation whether this edge has to be swapped to improve the connection structure. Figure 33 shows that an interior edge E_i from P_1 to P_2 is a candidate for swapping if, and only if, the sum of the interior angles of the two triangles at P_1 and P_2 is less than 180 degrees, because otherwise swapping E_i would result in overlapping triangles. Frey and Field[45] compute a so-called relaxation index R_i for E_i,

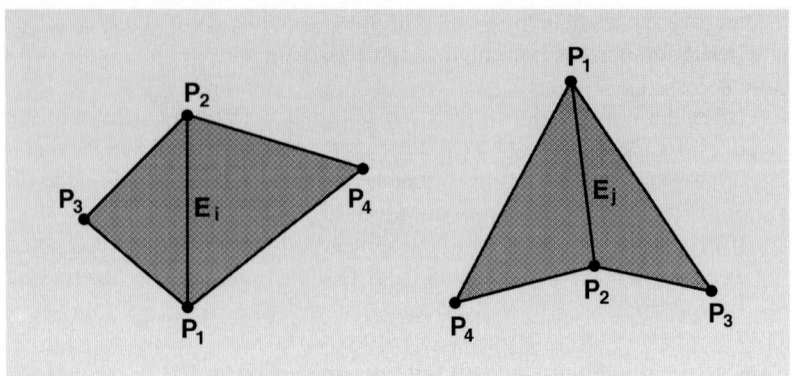

Figure 33. E_i is a candidate edge for swapping, whereas E_j is not.

depending on the degree of the nodes of the triangles sharing E_i. Let the degree of the four nodes be d_1, d_2, d_3 and d_4. Then R_i defined as:

$$R_i = (d_1 - 6) + (d_2 - 6) - (d_3 - 6) - (d_4 - 6) \qquad (10)$$
$$= d_1 + d_2 - d_3 - d_4$$

is a measure of the desirability to swap edge E_i. Swapping an edge for which $R_i > 2$, leads to a relaxation index closer to zero, and, according to this criterion, to a triangulation with a better connection structure.

Using [10] as a criterion, the occurrence of two 5-nodes is equally weighed as the occurrence of a single 4-node. In this way, the two configurations in Figure 34 would be considered equally desirable. However, it is clear that having a 4-node in the triangulation is less desirable than having two extra 5-nodes, since the occurrence of 4-nodes generally results in

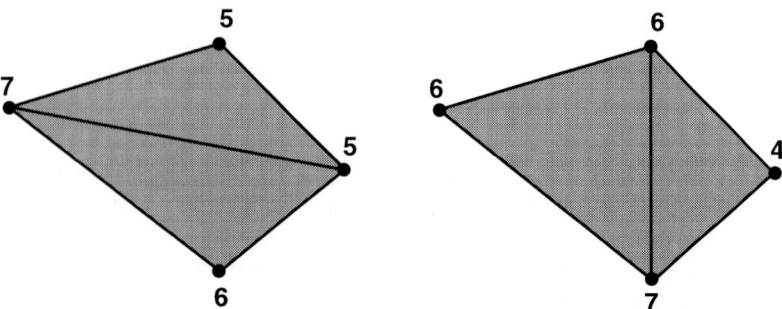

Figure 34. Two configurations that are equally desirable according to criterion [10]. The numbers at the nodes indicate their degree.

triangles with obtuse angles. In[46] a new relaxation index is defined, which stresses the importance of eliminating nodes with an extreme degree. R_i is now defined as

$$R_i = (d_1 - 6)^2 + (d_2 - 6)^2 + (d_3 - 6)^2 + (d_4 - 6)^2 \tag{11}$$

and an edge E_i will be swapped if the relaxation index R_i' after swapping is closer to zero than R_i.

The algorithm for improving a triangulation that uses this criterion, first swaps edges for configurations in the triangulation that give the most significant improvements in the connection structure. Following this, the remaining undesirable configurations are eliminated. To this end, first each edge for which $R_i - R_i' \geq T$ is swapped, where T is a positive threshold value. This process is then repeated for successively lower thresholds, until $T = 1$, or even $T = 0$, the neutral threshold. Nodes on the boundary of the triangulation require a special treatment for the computation of their degree; two approaches are described in.[45,46]

After edge swapping, a triangulation with an improved connection structure, but with many badly-shaped triangles will usually result. The second step of the relaxation algorithm is Laplacian smoothing of the triangulation, to improve the angles of the triangles.

In 3D, face swapping can be used to improve the shape of the tetrahedrons in a mesh. In[47] a method based on face swapping is used to improve a Delaunay tetrahedrization. It is observed that a Delaunay tetrahedrization does not seem to satisfy any global or localoptimality criterion. In practice, a Delaunay tetrahedrization will often contain badly-shaped elements.[14] To improve the tetrahedrization, the local max-min solid angle criterion is used. This criterion is satisfied if, for all groups of two or three adjacent tetrahedrons in certain configurations, the minimum of the solid angles at the nodes of the tetrahedrons is maximized over the possible tetrahedrizations of the nodes in this group.[47] The results indicate that significant improvements can be achieved in this way.

An interesting and simple algorithm that changes both the geometry and topology of a mesh consisting of Delaunay triangles is described in.[48] A disadvantage of Laplacian smoothing is that it does not always preserve the Delaunay properties of a triangulation. Laplacian-Delaunay smoothing does preserve the Delaunay properties by locally adapting the triangulation after each repositioning of an interior node. The algorithm works as follows:

– First, compute the new position P_{new} according to formula [9].
– Then, if moving the node to the new position preserves the Delaunay triangulation, simply update the position. Otherwise, delete the node, locally retriangulate to maintain a Delaunay triangulation, and insert a new node at the newly computed position.

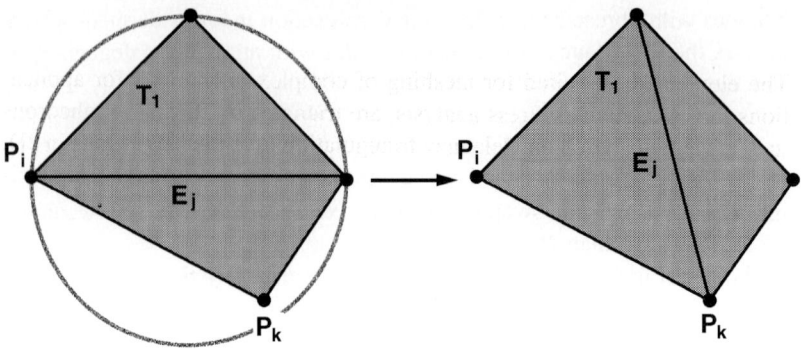

Figure 35. P_k falls within the circumscribing circle of T_1, and thus E_j has to be swapped.

A variant of this algorithm, described in,[46] does not involve the deletion and insertion of nodes, and is therefore more efficient. The algorithm consists of the following steps:

- First, compute the new position P_{new} for an interior node P_i using formula [9].
- Then, adapt the triangulation using an algorithm based on edge swapping:
  - For each interior edge E_j adjacent to the repositioned node P_i, test the triangles sharing E_j to check whether they are Delaunay triangles. The test for one triangle sharing E_j is done by determining whether its circumscribing circle contains a node of the other triangle sharing E_j (Figure 35).
  - If one of the triangles sharing E_j is no Delaunay triangle then swap E_j to maximize the smallest angle of the two triangles (Figure 35).
  - After swapping an edge, the four other edges used by the two triangles become candidates for swapping, since swapping an edge changes the angles of the triangles.
  - Apply the edge swapping test recursively to all these edges.

Using this algorithm, usually three to five iterations are enough to get a significant improvement in the shape of the triangles. The minimum and maximum angles in the triangulation can be expected to be significantly better with Laplacian-Delaunay smoothing than with Laplacian smoothing only. However, it is our experience[46] that applying mesh relaxation gives better results than Laplacian-Delaunay smoothing.

3.9. CONCLUSIONS

The elements best suited for meshing of complex geometries for applications in CAD, such as stress analysis, are triangles in 2D and tetrahedrons in 3D. Methods based on Delaunay triangulation give the best results in 2D, in the sense that the sum of the smallest angles over all triangles is maximized. The triangulation can be further improved using mesh improvement techniques.

For 3D, it is less clear which method results in the best tetrahedrization. All methods presented here have drawbacks. The algorithms based on Delaunay tetrahedrization are reasonably simple, but can create badly-shaped elements and suffer from numerical problems. However, mesh improvement techniques can be used to improve the created mesh.

Advancing front methods lack a theoretical basis for node creation/selection when forming a new element. A further disadvantage is the requirement to check the geometry of a newly formed element for non-intersection against the edges or triangles of the advancing front. Moreover, it is certainly not guaranteed that well-shaped elements are formed when at the end the front folds onto itself. All these disadvantages are rather severe when using the advancing front method to generate 3D meshes.

The use of methods that decompose or subdivide the object into simpler pieces is an option to localize computations. However, an important disadvantage of the spatial decomposition techniques is that the geometry of an object within a cell is not guaranteed to be less complex than that of the object itself. Of the methods discussed here, the medial axis subdivision method is the most interesting, because the resulting meshes are well structured and have a high quality. Unfortunately, there is no 3D version of this method available.

The 'standard' geometric model used in mesh generation is the boundary representation, mostly with curved faces approximated by planar faces. An alternative is to use an exact CSG representation of the object. Unfortunately, the mesh generation method developed for such a model is not yet reliable enough.

Feature modelling is a relatively new development in CAD. Features can be useful in the mesh generation process, because they provide a means of linking functional information to an object. Boundary conditions can be linked directly to features, and features can make relevant regions of an object explicit, which can be used during mesh generation. So feature modelling might play an important role in the future of mesh generation.

Several techniques have been developed that aim to improve the shape of the elements of an existing mesh. As long as there are no mesh generation methods that produce an 'optimal' mesh, these techniques will play an important role in the mesh generation process.

References

1. Strang, G. and G.J. Fix, 1973, *An analysis of the finite element method*. Prentice-Hall, Englewood Cliffs, NJ.
2. Zienkiewicz. O.C., 1977, *The finite element method*. McGraw-Hill, London, 3rd edition.
3. Hughes, T.J.R., 1987, *The finite element method*. Prentice-Hall, Englewood Cliffs, NJ.
4. Klein, R. and W. Straßer, 1995, *Proceedings Third Symposium on Solid Modeling and Applications*, 431–440.
5. Babuska, I. and A.K. Aziz, 1976, *SIAM Journal on Numerical Analysis*, **13**(2), 214–226.
6. Szabo, B.A., 1988, *Communications in Applied Numerical Methods*, **4**(3), 393–400.
7. Cavendish, J.C., 1974, *International Journal for Numerical Methods in Engineering*, **8**(4), 679–696.
8. Lo, S.H., 1985, *International Journal for Numerical Methods in Engineering*, **21**(8), 1403–1426.
9. Shimada, K. and D.C. Gossard, 1992, *IFIP Transactions B [Applications in Technology]*, **B-3**, 411–420.
10. Lawson, C.L., 1977, Software for C1 surface interpolation. In J.R. Rice, editor, *Mathematical software III*, pp. 161–194. Academic Press, New York.
11. Sibson, R., 1978, *The Computer Journal*, **21**(3), 243–245.
12. Watson, D.F., 1981, *The Computer Journal*, **24**(2), 167–172.
13. De Floriani, L., B. Falcidieno and C. Pienovi, 1985, *Computer Vision, Graphics and Image Processing*, **32**, 127–140.
14. Cavendish, J.C., D.A. Field and W.H. Frey, 1985, *International Journal for Numerical Methods in Engineering*, **21**(2), 329–347.
15. Peraire, J., M. Vahdati, K. Morgan and O.C. Zienkiewicz, 1987, *Journal of Computational Physics*, **72**(2), 449–467.
16. Peraire, J., J. Peiro, M. Formaggio, K. Morgan and O.C. Zienkiewicz, 1988, *International Journal for Numerical Methods in Engineering*, **26**(10), 2135–2159.
17. Löhner, R. and P. Parikh, 1988, *International Journal for Numerical Methods in Fluids*, **8**(10), 1135–1149.
18. Frey, W.H., 1987, *International Journal for Numerical Methods in Engineering*, **24**(11), 2183–2200.
19. Woo, T.C. and T. Thomasma, 1984, *Computers and Structures*, **18**(2), 333–342.
20. Wördenweber, B., 1984, *Computer-Aided Design*, **16**(5), 285–291.
21. Wördenweber, B., 1984, Finite-element analysis for the naive user. In M.S. Picket and J.W. Boyse, editors, *Solid modeling by computers: From theory to applications*. Plenum Press, New York.
22. Shephard, M.S. and M.K. Georges, 1991, *International Journal for Numerical Methods in Engineering*, **32**(4), 709–749.
23. Samet, H., 1990, *The design and analysis of spatial data structures*. Addison Wesley, Reading, MA.
24. Yerry, M.A. and M.S. Shephard, 1983, *IEEE Computer Graphics and Applications*, **3**(1), 39–46.
25. Baehmann, P.L., S.L. Wittchen, M.S. Shephard, K.R. Grice and M.A. Yerry, 1987, *International Journal for Numerical Methods in Engineering*, **24**(6), 1043–1078.
26. Bern, M. and D. Eppstein, 1992, Mesh generation and optimal triangulation. In F.K. Hwang and D. Du, editors, *Euclidean geometry and the computer*. World Scientific, Singapore.
27. Bern, M., D. Eppstein and J.R. Gilbert, 1990, *Proceedings 31st IEEE symposium on foundations of computer science*. IEEE, New York.
28. Baker, B.S., E. Grosse and C.S. Rafferty, 1988, *Discrete and Computational Geometry*, **3**, 147–168.

29. Melissaratos, E. and D. Souvaine, 1991, Coping with inconsistencies: a new approach to produce quality triangulations of polygonal domains with holes for the finite element method, Report LCSR-TR-163. Rutgers University, Laboratory for Computer Science Research, New Brunswick, NJ.

30. Yerry, M.A. and M.S. Shephard, 1984, *International Journal for Numerical Methods in Engineering*, **20**(11), 1965–1990.

31. Schroeder, W.J. and M.S. Shephard, 1988, *International Journal for Numerical Methods in Engineering*, **26**(11), 2503–2515.

32. Schroeder, W.J. and M.S. Shephard, 1990, *International Journal for Numerical Methods in Engineering*, **29**(1), 37–55.

33. Schroeder, W.J., and M.S. Shephard, 1989, *Engineering with Computers*, **5**(3-4), 177–193.

34. Field. D.A. and W.D. Smith, 1991, *International Journal for Numerical Methods in Engineering*, **31**(3), 413–425.

35. Tam, T.K.H. and C.G. Armstrong, 1991, *Advances in Engineering Software and Workstations*, **13**(5-6), 313–324.

36. Lee, Y.T., A. De Pennington and N.K. Shaw, 1984, *ACM Transactions on Graphics*, **3**(4), 287–311.

37. Yerry, M.A., and M.S. Shephard, 1984, *International Journal for Numerical Methods in Engineering*, **20**(11), 1965–1990.

38. Perucchio, R., M. Saxena and A. Kela, 1989, *International Journal for Numerical Methods in Engineering*, **28**(11), 2469–2501.

39. Boender, E., 1992, *Finite element mesh generation from CSG models*. Delft University of Technology, Delft.

40. Boender, E., W.F. Bronsvoort and F.H. Post, 1994, *Computer-Aided Design*, **26**(5), 379–391.

41. Bronsvoort. W.F. and F.W. Jansen, 1993, *Computers in Industry*, **21**(1), 61–86.

42. Unruh, V. and D.C. Anderson, 1992, *Engineering with Computers*, **8**, 1–12.

43. Cavendish, J.C., W.H. Frey and S.P. Marin, 1991, *Advances in Engineering Software and Workstations*, **13**(5-6), 226–237.

44. Edelsbrunner, H., T.S. Tan and R. Waupotitisch, 1990, Proceedings of the 6th annual symposium on computational geometry.

45. Frey, W.H. and D.A. Field, 1991, *International Journal for Numerical Methods in Engineering*, **31**(6), 1121–1131.

46. Boender, E., 1994, *Communications in Applied Numerical Methods*, **10**, 773–783.

47. Joe, B., 1991, *International Journal for Numerical Methods in Engineering*, **31**(5), 987–997.

48. Field, D.A., 1988, *Communications in Applied Numerical Methods*, **4**(6), 709–712.

4 AN ASYMPTOTIC-NUMERICAL METHOD FOR NONLINEAR VIBRATIONS OF ELASTIC STRUCTURES

L. AZRAR[1], B. COCHELIN[2], N. DAMIL[3] and M. POTIER-FERRY[4]

[1] *Département de Mathématiques, Faculté des Sciences et Techniques de Tanger, Université Abdelmalek Essaadi, BP 416 Tanger, Morocco or Laboratoire d'Études et de Recherches en Mathématiques Appliquées, E.G.T. École Mohammadia d'Ingénieurs, Université Mohammed V, BP 765 Agdal, Rabat, Morocco*

[2] *Institut Méditerranéen de Technologie, E.S.M.2 Technopôle de Chateau-Gombert, 13451 Marseille cedex 20, France*

[3] *Laboratoire de calcul Scientifique en Mécanique, Faculté des Sciences Ben M'Sik, Université Hassan II, Casablanca, Morocco*

[4] *Laboratoire de Physique et Mécanique des Matériaux, U.R.A. CNRS 1215, I.S.G.M.P. Université de Metz, Ile du Saulcy, 57045 Metz Cedex, France*

4.1. INTRODUCTION

The dynamic behaviour of thin elastic structures is governed by nonlinear partial differential equations. Generally, explicit solutions of these equations are not available in the literature owing to the difficulty of the mathematical treatment. Analytical solutions exist only for simple cases and therefore, numerical methods are necessary when considering more complex cases. The practical approach is to apply various approximate methods to certain special cases and this appears to be the only means available to gain an understanding of the systems in the nonlinear range. Considerable research

103

efforts have been devoted to obtain approximate solutions for non-linear response of thin elastic structures like beams, plates and shells. Free and forced vibrations of undamped thin elastic structures are of interest in the stability analysis. The amplitude-frequency relation, i.e. the backbone curve, is of primary importance to investigate the resonance phenomena.

The methods to study resonant vibrations of geometrically nonlinear elastic structures can be broadly divided into three groups. The assumption that the nonlinear mode shape is the same as the linear one is the basis of the first method. The governing partial differential equations can be treated by using Galerkin's method. This helps to reduce the governing dynamic equation to a single nonlinear ordinary differential equation in time of Duffing type. The latter equation is treated by a harmonic approximation or by elliptic functions.[1–5] The nonlinear vibration of laminated anisotropic plates is studied by a combination of quadratic and cubic terms in Duffing equation.[9,10] The most comprehensive work on the geometrically nonlinear analysis of both static and dynamic behaviour of thin elastic plates is Chia's standard book.[8] A survey of literature on nonlinear vibrations of plates can be found in.[6,7]

In the second approach, the dependence of the solution in time is assumed to be harmonic. Then, one can use a harmonic balance method to obtain a nonlinear boundary value problem in the space variable. This technique is largely used because it permits to transform the nonlinear dynamic problem into a nonlinear static one. This allows to work with the same numerical methods as in static. The classical finite element method can be used for an accurate solution of complex engineering problems. Most studies in nonlinear vibrations of structures using this approach have been done by combining the finite element method and linearising procedures.[11–16] As clearly presented in,[12,15,16] this procedure leads to an iterative linear eigenvalue problem. Without linearising functions the problem of nonlinear free vibrations is represented as a nonlinear eigenvalue problem which can be solved by an incremental iterative procedure.[17] Different authors use the reduced basis technique or the Ritz method.[18–20] The dynamic behaviour of plates at large vibration amplitudes was examined both theoretically and experimentally.[21–23] A semi analytical approach has been proposed by Benamar and White[24] and its application for forced nonlinear vibrations of beams has recently done.[25]

In the third approach one determines an approximate solution of nonlinear dynamic problems by treating the continuous problem directly without those simplifying assumptions. Different methodological approaches can be used like the method of multiple scales. For a comprehensive review of the literature, we refer the reader to Nayfeh *et al.*'s works.[26–28] The solution is generally expanded into a multiple Fourier series containing a number of frequencies. Theoretically there is no limitation to the number of time scales used, although it is obvious that the cost in implementation on computer will increase rapidly with the increasing number of frequencies and Fourier terms.[29] Other methods like symplectic integration or alternating frequency-time domain can be used.[30,31] The complete (tempero-spatial) problem in time and space has to be solved without any simplification. Remembering what are the difficulties to solve numerically some static or stationary problems, one understands that such methods will not be workable before long in the practice. Thus, it is relevant to look for simpler techniques for solving time-dependent problems.

The harmonic balance method is well known for highly nonlinear systems and largely used for nonlinear vibrations of elastic structures. Using this approach we get a nonlinear differential problem that is similar to a static one. The resolution of this problem permits to get the nonlinear modes of vibration and the backbone curves which are simple dynamic characteristics of the system. In general the computation methods used to solve this problem are the incremental iterative methods. The most popular one is a Newton-Raphson method associated with control parameters. The principle is to follow the nonlinear solution path in a stepwise manner, via a sequence of linearizations and some iterations to achieve equilibrium. With a proper parametrisation of the branch, such algorithms are successful in determining a complete shape of solution that will be represented point by point. But this requires long computation times as compared to a linear problem and it is difficult to automatize such step by step procedures.

An alternative approach corresponds to analytical representation techniques such as the perturbation methods. According to these representation techniques, the solution of the full problem is represented under the form of power series with respect to a path parameter. The principle of these methods is to determine some terms of the series by solving a recursive set of linear problems. In contrast to the incremental-iterative methods, perturbation methods have received much less attention from computational community. There are two main explanations for the poor success of such methods in computational context. The first one is the growing complexity of the right-hand sides of the linear problems obtained. The second is the thought that the analytical representation is valid only for very small values of the perturbation parameter. So, these methods are usually seen more as a theoretical framework for qualitative analysis than as a numerical tool to furnish accurate qualitative results.

An Asymptotic-Numerical Method based on perturbation techniques and finite element method has been developed for nonlinear problems. This method permits to remove some of the previous difficulties in the classical perturbation methods. It has been proposed by Damil and Potier-Ferry

(1990)[32] for computing perturbed bifurcation and applied in Azrar *et al.* (1993)[33–36] for computing the post-buckling behaviour of elastic plates and shells. Next, it has been extended to some nonlinear elastostatic problems in Cochelin *et al.* (1994).[37–38] These works have brought out the essential features that affect the practicability of *the coupling of perturbation methods and finite element methods*. The principle of this method is to represent the unknowns (displacements, load parameters, frequency,...) by a power series expansion with respect to a control parameter. By introducing the expansion into the governing equation, the nonlinear problem is transformed into a sequence of simple linear problems which we can solve by a classical finite element method. Hence, a large number of terms of the series can be numerically computed.

The main objective of this paper is to present the development of the Asymptotic-Numerical Method for nonlinear dynamic problems and its application for studying the nonlinear vibrations of elastic plates. The paper is organised as follows. First, the governing equations based on the Von-Karman's plate theory is presented. A discussion of the usual separation of variables which is rarely done in the literature is given. The use of harmonic balance method permits to obtain a simple operational formulation. The asymptotic-numerical method and its application for the backbone curve is described. Some details about relevant computation are given. Comprehensive numerical tests for nonlinear free vibration of plate are reported and discussed. A large part of the backbone curves are obtained. With this Asymptotic-Numerical Method we can easily compute a great number of terms of the asymptotic series. These terms reveal the important features of the nonlinear solutions and the remaining ones give small corrections. The essential work presented here is the adaptation of this method to nonlinear vibrations and the computation of the asymptotic series. However, the range of validity of the obtained solution is limited. Thus, to extract maximal information from computed terms of the series, some sophisticated techniques can be easily added. Methods to achieve these goals are presented in the static study in the last sections. Some reference results will be presented.

4.2. GOVERNING EQUATIONS

The modelling of the dynamic response of elastic structures introduces many additional considerations probably not anticipated from a static analysis. The inertial effects must be taken into account and the dynamic behaviour of the system must be treated as a function of time. In this chapter we shall treat the nonlinear vibrations of elastic plates by the harmonic balance method as an example of nonlinear dynamic analysis. Also we shall present the nonlinear elastostatic analysis of plates and shells. The differential equations of the motion governing the moderately large amplitude vibration of thin elastic plates can be obtained form the Von Karman's deflection theory of plates. The displacement variational principle of these equations leads to a cubic nonlinearity. The mixed stress-displacement approach, called Hellinger-Reissner principle leads to an only quadratic non linearity. Then in view of applying the expansion procedure, the mixed approach is preferred here since it leads to a simple algebra. After the expansion process, we shall come back to a displacement formulation and use the very classical displacement finite element method.

4.2.1. Formulation of the Potential Energy and Kinetic Energy

Let us note the displacement of the middle surface of the plate or the shell by u_1, u_2 and w, where u_α are the in-plane displacements and w the transverse displacement in the x_1, x_2, x_3 direction respectively. The reference $x_3 = 0$ may be chosen at an arbitrary location through the undeformed plate and the x_3 axis directed normally to the reference plane.

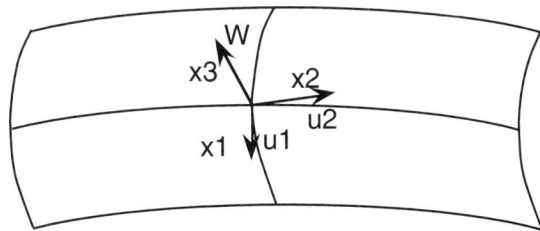

Reference configuration of the plate (or of the shell).

For thin structures, the Green-Lagrange strains are supposed to vary linearly through the thickness x_3 and will be noted by

$$\Gamma(u_\alpha, w) + x_3\ \kappa(u_\alpha, w)$$

with   (1)

$$\Gamma(u_\alpha, w) = \Gamma^L(u_\alpha, w) + \Gamma^{NL}(u_\alpha, w)$$

The tensors Γ and κ are the generalised membrane strain and bending strain. We separate the membrane Γ strain into a linear part Γ^L and non-linear part Γ^{NL}. From Von Karman deflection theory of plate the components of these generalised strain tensors are defined as:

$$\begin{cases} \Gamma^L_{\alpha\beta} = \frac{1}{2}(u_{\alpha\beta} + u_{\beta,\alpha}) \\ \Gamma^{NL}_{\alpha\beta} = \frac{1}{2}(w_{,\alpha} + w_{,\beta}) \\ \kappa_{\alpha\beta} = -w_{,\alpha\beta} \end{cases} \tag{2}$$

The bending strain κ is supposed to be linear with respect to the displacement (framework of moderate rotations). In the case of Donnell cylindrical shell of radius R we have the same formulas except for the following one

$$\Gamma^L_{2,2} = u_{2,2} - \frac{w}{R}$$

The membrane or in-plane forces N and moments M are assumed to be related to the strain and curvature by the constitutive relations:

$$\begin{cases} N = C_m : \Gamma \\ M = C_b : \kappa \end{cases} \tag{3}$$

where C_m and C_b are symmetric matrices of material properties. The more general situation with coupling matrix C_{mb} between membrane and flexion which is of great concern for unsymmetric composite laminates will not be presented here for simplicity.

The total strain energy expression of a plate (or of a shell) can be written as:

$$P(u_\alpha, w) = \frac{1}{2}\int_\Omega (\Gamma : C_m : \Gamma + \kappa : C_b : \kappa)\, d\Omega - \lambda\, P(u_\alpha, w) \tag{4}$$

where λ is the load parameter and λP is the work done by the applied loads. Since Γ is quadratic in u_α and w, $P(u_\alpha, w)$ is of the degree 4 with respect to u_α and w. To reduce the order of the nonlinearity we introduce an additional variable. This can be achieved by using the mixed Hellinger-Reissner functional:

$$H(u_\alpha, w, N) = \int_\Omega (N : \Gamma - \frac{1}{2}N : C_m^{-1} : N + \frac{1}{2}\kappa : C_b : \kappa)\, d\Omega - \lambda\, P(u_\alpha, w) \tag{5a}$$

where the unknown is the mixed (displacement-stre-ss) vector

$$U = {}^t[u_1, u_2, w, N] \tag{5b}$$

The mixed functional H is cubic in (u_1, u_2, w, N) and the variation δH yields to quadratic equations.

$$\delta H(u_\alpha, w, N) = \int_\Omega (N: \delta\Gamma + (\Gamma - N\,C_m^{-1}): \delta N + \kappa: C_b: \delta\kappa)\, d\Omega - \lambda\delta P(u_\alpha, w)$$

(6a)

The variational principle $\delta H(u_\alpha, w, N) = 0$ is equivalent to

$$\begin{cases} \delta P(u_\alpha, w) = 0 & \text{The virtual work principle} \\ N = C_m: \Gamma & \text{The constitutive equation} \end{cases}$$

(6b)

Since we shall work with the power series expansion it would be better to use the functional H which leads to a quadratic nonlinearity.

As for the kinetic energy T of a plate, we neglect the rotary inertia terms, which is the usual procedure for plate deflection. This gives the following functional:

$$T = \frac{1}{2}\int_\Omega \rho(\dot{u}_1^2 + \dot{u}_2^2 + \dot{w}^2)\, d\Omega$$

(7)

where ρ is the mass density and the overdot means the differentiation with respect to time.

4.2.2. Separation of Variables

In this section we present a discussion of the usual separation of variables and we develop a methodological approach for studying nonlinear forced vibration of plates in a simple way. Two types of simplified approaches can be found in the literature. In both cases, the deflection is assumed to be the product of a function of time and of a function of space

$$w(x, y, t) = w(x, y)\ \phi(t)$$

(8)

In the first approach, the mode shape $w(x, y)$ is prescribed and the amplitude is deduced from Hamilton's variational principle. In the second approach, the time dependence is assumed to be harmonic and the mode shape follows from the variational principle. In addition to the severe restriction (8), two assumptions are made here. First, the applied load is split into tangential loads proportional to a function $\lambda_1(t)$ and into normal loads proportional to an other function $\lambda_2(t)$.

$$\lambda P(u_\alpha, w) = \lambda_1(t)\ P_\alpha(u_\alpha) + \lambda_2(t)\ P_3(w)$$

(9)

Second, we limit ourselves in this subsection to a single flat Van Karman's plate in such a way that linearised equations can be uncoupled. Next, we also neglect the in-plane inertia terms, because their corresponding eigenfrequencies are very high. So the in-plane equations are static ones, their solution can be computed from the theorem of the potential energy and not from the Hamilton's principle. With account of (8–9) it can be written in the following form:

$$\int_\Omega \delta\Gamma^L : C_m : \Gamma^L(u_\alpha)\, d\Omega = \lambda_1(t)\mathcal{P}_\alpha(\delta u_\alpha) - \phi^2(t)\int_\Omega \delta\Gamma^L : C_m : \Gamma^{NL}(w)\, d\Omega \quad (10)$$

Because of the linearity of the latter equation with respect to u_α, it appears that the in-plane displacement can be split up as follows:

$$u_\alpha(x,y,t) = u_\alpha^1(x,y)\,\lambda_1(t) + u_\alpha^2(x,y)\,\phi^2(t) \quad (11)$$

Insertion of (8) and (11) into (2) and (3) leads to the formulation of the strain, curvature and the membrane tensors respectively as follows:

$$\Gamma^L(x,y,t) = \gamma_1^L(x,y)\,\lambda_1(t) + \gamma_2^L(x,y)\,\phi^2(t)$$

where

$$\gamma_\beta^L(x,y) = \left\{ \begin{array}{l} u_1^\beta(x,y)_{,x} \\ u_2^\beta(x,y)_{,y} \\ u_1^\beta(x,y)_{,y} + u_2^\beta(x,y)_{,x} \end{array} \right\} \quad (12)$$

$$\Gamma^{NL}(x,y,t) = \left\{ \begin{array}{l} \dfrac{1}{2}w_{,x}^2 \\ \dfrac{1}{2}w_{,y}^2 \\ w_{,x}\,w_{,y} \end{array} \right\} \phi^2(t) = \gamma^{NL}(x,y)\,\phi^2(t) \quad (13)$$

$$\Gamma(x,y,t) = \gamma_1^L(x,y)\,\lambda_1(t) + [\gamma_2^L(x,y) + \gamma^{NL}(x,y)]\,\phi^2(t) \quad (14)$$

$$K(x,y,t) = \left\{ \begin{array}{l} -w_{,xx} \\ -w_{,yy} \\ -2w_{,xy} \end{array} \right\} \phi(t) = K(x,y)\,\phi(t) \quad (15)$$

$$N(x,y,t) = N_1(x,y)\,\lambda_1(t) + N_2(x,y)\,\phi^2(t) \text{ where}$$

$$N_1(x,y) = [C_m]:\gamma_1^L(x,y) \text{ and } N_2(x,y) = [C_m]:\gamma_2^L(x,y) + \gamma^{NL}(x,y)] \quad (16)$$

After insertion of (12,...,16) into equation (5) we obtain the strain energy of plate as follows:

$$H(u_\alpha, w, N) = \lambda_1^2(t)\int_\Omega \left(N_1 : \gamma_1^L - \frac{1}{2}N_1 : C_m^{-1} : N_1 \right) d\Omega$$

$$+ \lambda_1(t)\phi^2(t)\int_\Omega \left[N_1 : \left(\gamma_2^L + \gamma^{NL} \right) + N_2 : \gamma_1^L - N_1 : C_m^{-1} : N_2 \right] d\Omega$$

$$+ \phi^4(t)\int_\Omega \left[N_2 : \left(\gamma_2^L + \gamma^{NL} \right) - \frac{1}{2}N_2 : C_m^{-1} : N_2 \right] d\Omega$$

$$+ \phi^2(t)\int_\Omega \frac{1}{2}\kappa : C_b : \kappa \, d\Omega - \lambda_1^2(t)\mathcal{P}_\alpha(u_\alpha^1)$$

$$- \lambda_1(t)\phi^2(t)\mathcal{P}_\alpha(u_\alpha^2) - \lambda_2(t)\phi(t)\mathcal{P}_3(w)$$

The kinetic energy when neglecting the in-plane inertia is given by:

$$T = \dot{\phi}(t)^2 \int_\Omega \frac{1}{2}\rho\, w^2\, d\Omega \tag{18}$$

At the moment, we assume that the plate is only subjected to transverse loading ($\lambda_1(t) = 0$). Following the in-plane equilibrium equations in (10), the adjusted separation of variables will be:

$$\begin{cases} u_1(x,y,t) = u_1(x,y)\,\phi^2(t) \\ u_2(x,y,t) = u_2(x,y)\,\phi^2(t) \\ w(x,y,t) = w(x,y)\,\phi(t) \end{cases} \tag{19}$$

For the sake of simplicity, we drop the fictitious indices. The strain tensor Γ, the constraint N and equation (14) can be written as:

$$\Gamma(x, y, t) = \gamma(x, y)\,\phi^2(t) \quad \text{and} \quad N(x, y, t) = N(x, y)\,\phi^2(t)$$

$$\int_{t1}^{t2}(H - T)\, dt = \int_{t1}^{t2}\phi^4(t)\, dt \int_\Omega \left(N : \gamma - \frac{1}{2}N : C_m^{-1} : N \right) d\Omega$$

$$+ \int_{t1}^{t2}\phi^2(t)\, dt \int_\Omega \frac{1}{2}\kappa : C_b : \kappa \, d\Omega \tag{21}$$

$$+ \int_{t1}^{t2}\phi(t)(-\lambda_2(t)\mathcal{P}_3(w))\, dt - \int_{t1}^{t2}\dot{\phi}(t)^2\, dt \int_\Omega \frac{1}{2}\rho\, w^2\, d\Omega$$

Let us assume that the modal displacemant is known, as presented in the first approach in the introduction. For instance, the assumed deflection $w(x, y)$ could be chosen as a linear vibration mode and the assumed in-plane displacement could be deduced from the latter by an equation like (10). Then we obtain:

$$\int_{t1}^{t2} (H - T)\, dt = \int_{t1}^{t2} \left[\phi^4(t)\frac{C1}{4} + \phi^2(t)\frac{C2}{2} - \lambda_2(t)\phi(t)C_3 - \dot{\phi}(t)^2\frac{C4}{2} \right] dt$$

(22)

where C_i ($i = 1,4$) are the constants corresponding to the integrals over Ω in equation (21). The Hamilton's principle $\delta\int_{t1}^{t2} (H - T)\, dt = 0$ leads to:

$$\int_{t1}^{t2} \{[\phi^3(t)C_1 + \phi(t)C_2 - \lambda_2(t)C_3]\delta\phi(t) - \dot{\phi}(t)\,\delta\dot{\phi}(t)C_4]\}\, dt \qquad (23)$$

Upon integration by parts of the last term and substitution of $\delta\phi(t_1) = \delta\phi(t_2) = 0$ we obtain the nonlinear ordinary differential equation called Duffing equation for the time function $\phi(t)$.

$$\ddot{\phi}(t)C_4 + \phi(t)C_2 + \phi^3(t)C_1 = \lambda_2(t)C_3 \qquad (24)$$

This equation is usually used to study the nonlinear free and forced transverse vibrations of plates.[8]

Now, we assume that the plate is subjected to the transverse and the lateral load. The displacement (u_1, u_1, w) is given by (8) and (11). The same development as presented before leads to the following integral:

$$\int_{t1}^{t2} (H - T)\, dt = \int_{t1}^{t2} \left[\lambda_1^2(t)D1 + \lambda_1(t)\phi^2(t)\frac{D2}{2} + \phi^4(t)\frac{D3}{4} + \phi^2(t)\frac{D4}{2} \right.$$
$$\left. - \dot{\phi}(t)^2\frac{D5}{2}\lambda_1^2(t)D6 - \lambda_1(t)\phi^2(t)D7 - \lambda_2(t)\phi(t)D8 \right] dt$$

(25)

where Di ($i = 1...5$) are the constants corresponding to the integrals over Ω in equation (17). $D6 = \mathcal{P}_\alpha(u_\alpha^1)$, $D7 = \mathcal{P}_\alpha(u_\alpha^2)$ and $D7 = \mathcal{P}_3(w)$. The Hamilton's principle leads to the following Mathieu equation:

$$\ddot{\phi}(t)D_5 + \phi(t)[D_4 + \lambda_1(t)D_2'] + \phi^3(t)D_3 = \lambda_2(t)D_8 \qquad (26)$$

where $D_2' = D_2 - 2D_7$.

The governing equations (24, 26) can also be formulated using the admissible displacement functions satisfying the boundary conditions by the application of Galerkin's method. The nonlinear free and forced vibrations of plates can be done by solving the last two differential equations. The resolution of those equations can be easily done by direct numerical integration methods or by perturbation methods. The main advantage of the prescribed formulation is that it permits to study in a simple way the forced vibrations of plates with various geometries and boundary conditions by using the finite element methods.

The formulation (17) permits one to study non periodic vibration by using a direct integration method in time and space. It is well known that the aforementioned integration method is very time consuming. To overcome that, one can expand $\phi(t)$ into a multiple Fourier series containing a number of frequencies in-commensurable with one another. Obviously, this form of solution is aperiodic in general. This expansion permits especially to study the periodic vibration when the frequency of the Fourier series are proportional. By this way, the nonlinear vibration with internal resonance can be obtained. Theoretically, there is no limitation to the number of time scales used. However, it is evident that the cost in implementation on computer will increase rapidly with the increasing number of frequencies and Fourier terms. That is why most of the authors limit themselves to one or to two harmonics for periodic vibrations.

4.3. HARMONIC BALANCE METHOD AND OPERATIONAL FORMULATION

In the present work, only periodic vibrations of an undamped system are considered. For a simple representation, we limit ourselves here to study a harmonic motion in a simple way. We neglect the in-plane load ($\lambda_1 = 0$) and we assume that the time function $\phi(t)$ and the displacement are given by:

$$\begin{cases} \phi(t) = \sin \omega t \\ u_\alpha(x,y,t) = u_\alpha(x,y) \sin^2 \omega t \\ w(x,y,t) = w(x,y) \sin \omega t \end{cases} \tag{27}$$

where ω is the natural frequency. To study the history of the actual solution corresponding to a period, we put the initial time $t_1 = 0$ and the final time $t_2 = \dfrac{2\pi}{\omega}$. In the present framework, we introduce the in-plane inertia terms. After integration of equation (21) over the time one obtains

$$\int_0^{\frac{2\pi}{\omega}} (H-T)\, dt = \frac{\pi}{\omega} \left\{ \frac{3}{4} \int_\Omega \left(N:\gamma - \frac{1}{2} N:C_m^{-1}:N \right) d\Omega + \frac{1}{2} \int_\Omega \kappa:C_b:\kappa\, d\Omega \right.$$

$$\left. - \lambda P(w) - \omega^2 \frac{1}{2} \rho \int_\Omega (u_1^2 + u_2^2 + w^2)\, d\Omega \right\} \tag{28}$$

where λ is a constant which represent the load parameter $-\lambda P(w) = P_3(w) \int_0^{\frac{2\pi}{\omega}} \phi(t) \lambda_2(t)\, dt$. Using the Hamilton's principle, the governing equation is given by:

$$\frac{3}{4} \int_\Omega (\delta\gamma:N + \delta N:\gamma - \delta N:C_m^{-1}:N)\, d\Omega + \int_\Omega \delta\kappa:C_b:\kappa\, d\Omega - \lambda P(\delta w)$$

$$- \omega^2 \rho \int_\Omega (u_1\delta u_1 + u_2\delta u_2 + w\delta w)\, d\Omega = 0 \tag{29}$$

This mixed variational principle can be used directly by using a mixed finite element method. To obtain a displacement formulation one can determine the membrane strain N as a function of displacement.

$$N(x,y) = [C_m]\,[\gamma(u_\alpha, w] = [C_m]\,[\gamma^1(u_\alpha) + \gamma^{n1}(w,w)]$$

The insertion of the strain N in equation (29) leads to a variational principle of cubic nonlinearity in displacement. After discretization by finite element method, one gets a nonlinear eigenvalue problem. This problem can be treated by a predictor-corrector method. Our purpose here is to solve the variational equation (29) by using an asymptotic-numerical method.

We introduce the following change of variables which represents a perturbation of the frequency parameter ω in the vicinity of a linear eigenfrequency ω_0 by

$$\omega^2 = \omega_0^2 + \varepsilon \tag{30}$$

Inserting (30) in the equilibrium equation (29) one get

$$\frac{3}{4} \int_\Omega [N:\delta\gamma^1 + (\gamma^1 - C_m^{-1}:N):\delta N]\, d\Omega + \int_\Omega \delta\kappa:C_b:\kappa\, d\Omega$$

$$- \lambda P(\delta w) - \omega_0^2 \rho \int_\Omega [u_1\delta u_1 + u_2\delta u_2 + w\delta w]\, d\Omega \tag{31}$$

$$- \varepsilon \rho \int_\Omega [u_1\delta u_1 + u_2\delta u_2 + w\delta w]\, d\Omega + \frac{3}{4} \int_\Omega [N:\delta\gamma^{n1} + \gamma^{n1}:\delta N]\, d\Omega = 0$$

In view of using an operational notation as presented in,[33-36] the governing equation (31) can be written as:

$$<L.U, \delta U> - \varepsilon <M.U, \delta U> + <Q(U, U), \delta U> = \lambda <\mathcal{F}, \delta U> \quad (32a)$$

where $U = {}^t[u_1, u_2, w, N]$ is the mixed vector and

$$<L.U, \delta U> = \frac{3}{4} \int_{\Omega} [N: \delta\gamma^1 + (\gamma^1 - C_m^{-1}: N): \delta N]\, d\Omega$$

$$+ \int_{\Omega} \delta\kappa: C_b: \kappa\, d\Omega - \omega_0^2 \rho \int_{\Omega} [u_1 \delta u_1 + u_2 \delta u_2 + w \delta w]\, d\Omega \quad (32b)$$

$$<M.U, \delta U> = \rho \int_{\Omega} [u_1 \delta u_1 + u_2 \delta u_2 + w \delta w]\, d\Omega \quad (32c)$$

$$<Q(U, U), \delta U> = \frac{3}{4} \int_{\Omega} [N: \delta\gamma^{n1} + \gamma^{n1}: \delta N]\, d\Omega \quad (32d)$$

$$<\mathcal{F}, \delta U> = \mathcal{P}(\delta w) \quad (32e)$$

The operators L and M are linear and Q is a quadratic one. The matrices corresponding to the operators L and M are the linear stiffness and the mass matrix respectively.

4.4. ASYMPTOTIC-NUMERICAL METHOD

The asymptotic numerical method has been used to study the buckling and post-buckling of elastostatic structures like beams, plates and shells.[32,37] Our aim is to use this method to study the nonlinear vibrations of elastic structures. We present here it's developments for the nonlinear free vibrations of thin plates. The unknowns are the mixed vector U and the frequency parameter ω. We assume that (U_0, ω_0) is a solution of equation (32a) and that in the vicinity of this point the branch can be represented by power series with respect to a path parameter "a", that will be defined later.

$$\begin{cases} U = U_0 + \sum_{p=1}^{+\infty} a^p U(p) \\ \varepsilon = \sum_{p=1}^{+\infty} a^p C(p) \end{cases} \quad (33)$$

Hence, $U(p)$ are mixed unknown vectors and $C(p)$ are unknown coefficients. By introducing equation (33) into equation (32a) and equating like powers of "a", we obtain the following set of linear mixed problems.

order 0 $L.U_{(0)} + Q(U_{(0)}, U_{(0)}) = \lambda \mathcal{F}$

order 1 $L_t.U_{(1)} = C_{(1)}M.U_{(0)}$

order 2 $L_t.U_{(2)} = C_{(2)}M.U_{(0)} + C_{(1)}M.U_{(1)} - Q(U_{(1)}, U_{(1)})$ (34)

...

order p $L_t.U_{(p)} = C_{(p)}M.U_{(0)} + \displaystyle\sum_{r=1}^{p-1} C_{(r)}M.U_{(p-r)} - \sum_{r=1}^{p-1} Q(U_{(r)}, U_{(p-r)})$

where the operator L_t is defined as $L_t(.) = L(.) + 2Q(U_{(0)}, .)$. The principle of the numerical method is to compute successively a number of vectors $U(p)$, coefficients $C(p)$ up to a given order n. This formulation is general. It permits one to study the free vibrations ($F = 0$), and forced vibrations ($F \neq 0$).

To study the elastostatic problems we use the same formulation with ($\omega = 0 = \varepsilon$) and we do the asymptotic expansion for the load parameter λ instead of ω. For more informations about the use of this method to study the elastostatic structural problems we refer the reader to.[32–37] In this paper we limit ourselves to the nonlinear free vibration analysis of plates. For that, we put $F = 0$ in the equation (32a). We begin by a formulation of a solution branch that bifurcates from the fundamental one, which modelizes the free vibration analysis.

4.5. POLYNOMIAL APPROXIMATION OF A BIFURCATING BRANCH

The backbone curve that determines the free vibration analysis of plates is a branch that bifurcates from the fundamental solution $U = 0$ at $\omega = \omega_0$ where ω_0 is the linear frequency of vibration. At the bifurcation point ($U = 0$, $\omega = \omega_0$) the tangent operator L_t (here $L_t = L$) is singular. Furthermore we assume that the kernel of this tangent operator is one-dimensional, what occurs generally. Using the Lyapunov-Schmidt reduction method and implicit function theorem it can be established that the unknown vector U and the frequency parameter ω can be expanded into integro-power series of a parameter "a" in the vicinity of the bifurcating point in the form of equation (33) with the following orthogonality condition:

$$< U_{(p)}, U_{(1)} > = 0 \qquad \text{if } p > 1 \qquad (35)$$

This orthogonality condition comes from Lyapunov-Schmidt reduction and the assumption that the kernel of the tangent operator L is one-dimensional and generated by the vector $U_{(1)}$. At each order p we have to determine the unknowns $U_{(p)}$ and $C_{(p)}$. The truncature of the series (33) at the order n yields to polynomials $\varepsilon(a, n)$ and $U(a, n)$ that we consider as approximations of the exact solution branch.

At order 1, we have to solve the following variational problem

$$< L.U_{(1)}, \delta U > = 0 \tag{36a}$$

The resolution of this problem gives the linear vibrations of plates. This allows one to determine the linear frequency ω_0 and the corresponding vibration mode $U_{(1)}$.

At order 2, we have to determine $C_{(1)}$ and $U_{(2)}$ by solving the following variational problem:

$$< L.U_{(2)}, \delta U > = C_{(1)} < M.U_{(1)}, \delta U > - < Q(U_{(1)}, U_{(1)}), \delta U >$$
$$< U_{(2)}, U_{(1)} > \; = 0 \tag{36b}$$

Since the operator L is self adjoint and using equation (36a), (ie $< L.U_{(2)}, U_{(1)} > = < U_{(2)}, L.U_{(1)} > = 0$), the coefficient $C_{(1)}$ can be easily computed by putting $\delta U = U_{(1)}$ in the variational equation (36b).

$$C_{(1)} = \frac{< Q(U_{(1)}, U_{(1)}), U_{(1)} >}{< M.U_{(1)}, U_{(1)} >} \tag{37a}$$

In this way, the coefficient $C_{(p-1)}$ is given by:

$$C_{(p-1)} = \frac{1}{< M.U_{(1)}, U_{(1)} >} \left\{ -\sum_{r=1}^{p-2} C_{(p)} < M.U_{(p-r)}, U_{(1)} > + \sum_{r=1}^{p-1} < Q(U_{(r)}, U_{(p-r)}), U_{(1)} > \right\} \tag{37b}$$

Remark that we compute the coefficient $C_{(p)}$ in terms of the vectors $U_{(q)}$, $q \leq p$ and $C_{(p)}$ takes place in the determination of $U_{(p+1)}$. To compute the vector $U_{(p)}$, the only difficulty is to construct the right-hand-sides which depends on $U_{(q)}$ and $C_{(q)}$, $q < p$. The pth problem can be written under the more compact form:

$$\begin{cases} < L.U_{(p)}, \delta U > = < F_{(p)}, \delta U > \quad \forall \delta U & \text{(38a)} \\ \\ < U_{(p)}, U_{(1)} > = 0 & \text{for } p > 1 \quad \text{(38b)} \end{cases}$$

The operator $<L(.), \delta U>$ is defined by (32b) and the R.H.S. is given by:

$$< F_{(p)}, \delta U >= \int_{\Omega} \left(F^U(p)\, \delta U + \frac{3}{4} F^w_\alpha(p)\, \delta w,_\alpha + \frac{3}{4} F^N_{\alpha\beta}(p)\, \delta N_{\alpha\beta} \right) d\Omega$$

(39a)

where

$$F^U(p)\, \delta U = \sum_{r=1}^{p-1} C(r)\rho[u_1(p-r)\,\delta u_1 + u_2(p-r)\,\delta u_1 + w(p-r)\delta w] \quad (39b)$$

$$F^w_\alpha(p) = -\sum_{r=1}^{p-1} N_{\alpha\beta}(r)w,_\beta(p-r) \tag{39c}$$

$$F^N_{\alpha\beta}(p) = -\sum_{r=1}^{p-1} \frac{1}{2} w,_\alpha(r)w,_\beta(p-r) \tag{39d}$$

Let us recall that the unknown vector $U_{(p)}$ includes not only the displacements u_1, u_2 and w but also the membrane stress resultant $N_{\alpha\beta}$ which had been introduced to reduce the order of the nonlinearity to a quadratic one in the governing equation. As a consequence, all the linear problems (38) are mixed.

4.5.1. Transformation of the Linear Mixed Problems into Displacement Problems

In order to use classical F.E.M., we shall now transform (38) into a pure displacement problem and a pseudo-constitutive equation which gives the pth term of the resultant stress $N(p)$. To get $N(p)$ at order p, we put $\delta u_\alpha = \delta w = 0$ and $\delta N_{\alpha\beta} \neq 0$ in the equation (38) that yields to:

$$N(p) = [C_m]\ [\gamma^l(p) - F^N(p)] \tag{40}$$

The resultant stress $N(p)$ is given as a function of displacement. The insertion of (40) into (38, 39) leads to a pure displacement problem as follows: Let us note the displacement vector $\bar{U}$ by:

$$\bar{U} = \begin{Bmatrix} u_1 \\ u_2 \\ w \end{Bmatrix} \tag{41}$$

$$\begin{cases} <\overline{L}.\overline{U}(p), \delta\overline{U}> = <\overline{F}, \delta\overline{U}> \quad \forall \delta\overline{U} & \text{(42a)} \\[2mm] <\overline{U}(p), \overline{U}(1)> = 0 \quad \text{for } p > 1 & \text{(42b)} \end{cases}$$

with

$$\begin{aligned} <\overline{L}.\overline{U}(p), \delta\overline{U}> = &\frac{3}{4}\int_\Omega \gamma^1 : C_m : \delta\gamma^1 d\Omega + \int_\Omega \kappa : C_b : \delta\kappa \, d\Omega \\ &- \omega_0^2 \rho \int_\Omega (u_1 \delta u_1 + u_2 \delta u_2 + w\delta w) \, d\Omega \end{aligned} \qquad \text{(43a)}$$

$$<\overline{F}(p), \delta\overline{U}> = \int_\Omega \left(F^U(p)\delta U + \frac{3}{4}F_\alpha^w(p)\delta w,_\alpha + \frac{3}{4}C_m : F_{\alpha\beta}^N(p):\delta\gamma^1 \right) d\Omega \quad \text{(43b)}$$

The operator $<\overline{L}.\overline{U}(p), \delta\overline{U}>$ is similar to an operator of dynamical stiffness and it involves the elastic bending stiffness, the membrane stiffness and the mass operator. So, the mixed problem (38) has been replaced by the displacement problem (42) and the constitutive equation (40) gives the stress. Finally, N is a convenient additional variable to reduce the degree in the equation and to make easier the expansion procedure. Once the nonlinear problem is transformed into a set of linear problems, the stress is eliminated and a classical F.E.M. is used to solve the linear problems.

4.5.2. Computation Procedures of $U(p)$ and $C(p)$ with Classical F.E.M.

Before computing the vectors $\overline{U}_{(p)}$ and the coefficients $C(p)$ that give the backbone curve, we must determine the eigenmode and eigenfrequency characterising the linear vibrations. For that we have to solve by F.E.M. equation (36a), written in displacement form as follows:

$$<\overline{L}.\overline{U}_{(1)}, \delta\overline{U}> = 0. \qquad \text{(44a)}$$

With the classical notation of computational mechanics[39] the discretization of this problem leads to an eigenvalue problem in the form:

$$[Ke - \omega_0^2 M][\overline{U}_{(1)}] = 0 \qquad \text{(44b)}$$

where $[Ke]$ is the elastic stiffness matrix and $[M]$ is the mass matrix. 'The resolution of this problem gives the linear modes and frequencies of vibration. We assume that $[\overline{U}_{(1)}]$ is the first mode of vibration and ω_0 its corresponding frequency.

Computation of $\overline{U}(p)$

The discretization of the problem (42) reads

$$[Ke - \omega_0^2 M][\overline{U}_{(p)}] = [\overline{F}(p)] \tag{45}$$

where $[\overline{U}_{(p)}]$ is the vector of the modal displacement at order p. $[Ke - \omega_0^2 M]$ is the tangent stiffness matrix at a bifurcation point which is singular. The orthogonality condition (42b) between $\overline{U}_{(p)}$ and $\overline{U}_{(1)}$ should be added to (45) in order to get an invertible problem. After discretization, this condition reads:

$$[\overline{U}_{(1)}]^t . [P] . [\overline{U}_{(p)}] = 0. \tag{46}$$

where P is a positive-definite matrix associated with the scalar product (42b). Here we choose to take $[P] = [Ke]$ to have an energy oriented scalar product. So, after relaxation of (45) by equation (46) one gets an invertible problem

$$\left[\begin{array}{c|c} Ke - \omega_0^2 M & \overline{U}^* \\ \hline \overline{U}^{*t} & 0 \end{array} \right] \left[\begin{array}{c} \overline{U}_{(p)} \\ \hline k \end{array} \right] = \left[\begin{array}{c} \overline{F}(p) \\ \hline 0 \end{array} \right] \tag{47}$$

where $[\overline{U}^*] = [Ke][\overline{U}_{(1)}]$ and k is a Lagrange multiplier. Note that the stiffness matrix is the same for all linear problems. Hence, we perform a Crout decomposition once for all. The right hand side vector $\overline{F}(p)$ depends on the already calculated vectors $U(q)$ and coefficients $C(q)$ for $q < p$. The assembling of $\overline{F}(p)$ is very similar to the assembling of a residual vector in a Newton-Raphson scheme and requires about the same computation time. The computation of the coefficients $C(p)$ and the membrane stress $N_{\alpha\beta}(p)$ can be judiciously performed during the assembling of the R.H.S. $\overline{F}$ so that almost no additional computation time is required for these quantities.

In summary, we say that this asymptotic-numerical method requires about the same computation time as a modified Newton-Raphson step with $(n-1)$ iterations without reactualization of the stiffness matrix. The computation time of some vectors $V(p)$ in static case with elastic analysis, buckling analysis and Newton-Raphson steps have been done in Table 1.[36] For large number of d.o.f., most of the computating time is spent in the Crout decomposition. That is why the computation of many terms requires only 50% of additional computation time with respect to a linear analysis.

Table 1. Computing time of the asymptotic-Numerical Method in static case. Comparison between linear elastic analysis, linear buckling analysis and modified Newton-Raphson steps.[36]

	Number of d.o.f: 726		Number of d.o.f: 4812	
	Time in second (Workstation)	Ratio t/t elastic	Time in sec.	Ratio t/t elast.
Crout decompo. of $K = L^t DL$	24	0.4	223	0.9
Linear elastic analysis	60	1	247	1
Modified Newton-Raphson				
5 iterat.	118	1.96	291	1.17
Linear buckling analysis	334	5.56	1870	7.57
Asymptotic-Numerical Method				
Order2	85	1.41	270	1.09
Order5	110	1.83	293	1.18
Order10	169	2.81	333	1.34
Order15	242	4.03	379	1.53

4.6. NUMERICAL RESULTS

To bring out the effectiveness and reliability of this method, numerical results of nonlinear free vibration of thin elastic plates with various geometries and boundary conditions will be presented. Initially, the linear eigenvalue problem (44) is solved by using F.E.M. Due to the symmetry, only one-quarter of the plate is modelled with triangular shell elements D.K.T. which has three nodes and five d.o.f. per node (u_1, u_2, w, θ_1, θ_2).[39,40] This investigation gives the linear frequency ω_0 and the associated mode shape $\overline{U}_{(1)}$ which correspond to linear vibrations of plates. To obtain the nonlinear frequency and nonlinear mode shape, the analysis presented above suggests the resolution of the linear systems (47) for computing the displacement vectors $\overline{U}_{(p)} = {}^t\{u_1(p), u_2(p), w(p)\}$, where w is the deflection and u_1 and u_2 are the in-plane displacements. The coefficients $C(p)$ constituting the frequency are computed by the algebraic equation (37). The truncation of the series (33) at order n gives:

$$\begin{cases} \overline{U}(p) = a\,\overline{U}(1) + a^2\,\overline{U}(2) + a^3\,\overline{U}(3) + \dots + a^n\,\overline{U}(n) \\ \omega^2 = \omega_0^2 + a\,C(1) + a^2\,C(2) + a^3\,C(3) + \dots + a^n\,C(n) \end{cases} \quad (48)$$

These series permit to obtain the frequency (or the period) as a function of the displacement at desired point of the plate. All these asymptotic results are compared with published results. A discussion of numerical results for the backbone curves is presented for plates with various shapes and boundary conditions.

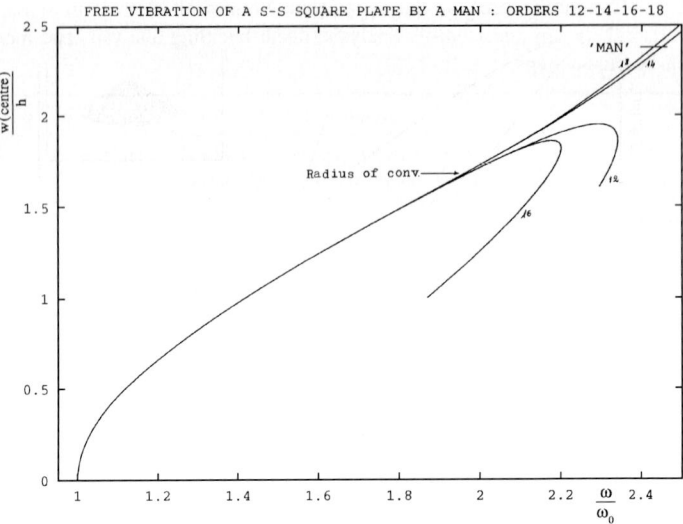

Figure 1. Displacement-frequency curve for a simply supported isotropic square plate: w(centre)/h versus the ratio ω/ω_0 at different truncatures of the asymptotique numerical solution. Orders 12–14–16–18.

    In Figure 1, we present the ratio $\dfrac{w(\text{centre})}{h}$ versus the ratio $\dfrac{\omega}{\omega_0}$ for different orders of truncatures 12–14–16–18 of the series (48). w(centre) is the deflection at the centre of the plate and h is the thickness.

    The frequency-displacement curve obtained for the plate problem are very characteristic of polynomial approximations. Indeed, for small values of the parameter "a", the Asymptotic-Numerical solutions (48) coincide quite perfectly until a critical value of "a". Beyond this value the truncated polynomial series separates from each other and diverges. Obviously, this critical value is *the radius of convergence* of the series (48) and it is clearly defined in all studied cases. The limitation of the validity of the solution is not an handicap of this method for two main reasons. First, the zone of convergence is sufficiently large. As presented in Figure 1, we obtain a large part of the backbone curve characterising the nonlinear vibrations of a simply supported square plate until $w \cong 2w_0$ and w(centre) $\cong 1.7h$. Second, the range of validity of the asymptotic solutions can be easily increased by using some shrewd strategies like Pade approximants[37] or the continuation techniques.[38] The effectiveness of these procedures will be presented in the following sections.

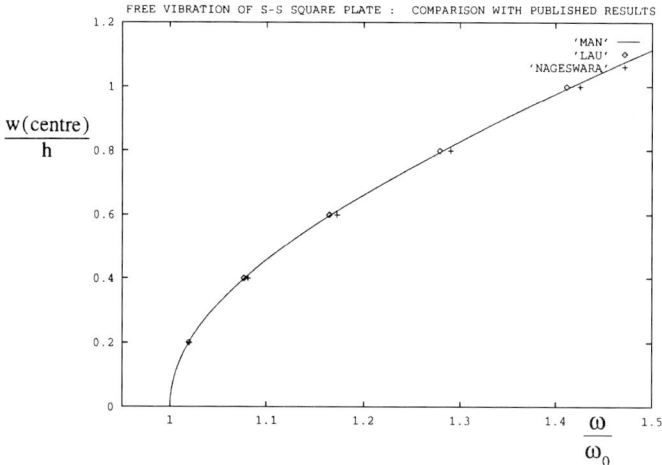

Figure 2. Displacement at the centre versus frequency ratio of a simply supported isotropic square plate ($L/h = 100$, $v = 0.3$). Comparison with results of Lau *et al.* [41] ◇ and Nageswara *et al.* [9] +.

Figures 2 and 3 show a plot of the ratio $\dfrac{T_{NL}}{T_L} = \dfrac{\omega_0}{\omega}$ versus the ratio w(centre)/h of simply supported and clamped square plates. A comparison of our results and those of other authors,[9,22,41] is given in the case of simply

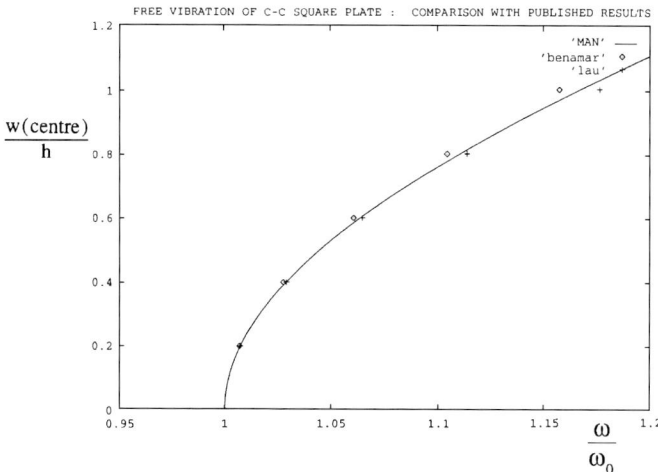

Figure 3. Displacement at the centre versus frequency ratio of a clamped-clamped isotropic square plate ($L/h = 100$, $v = 0.3$). Comparison with results of Lau *et al.* [41] + and Benamar *et al.* [22] ◇.

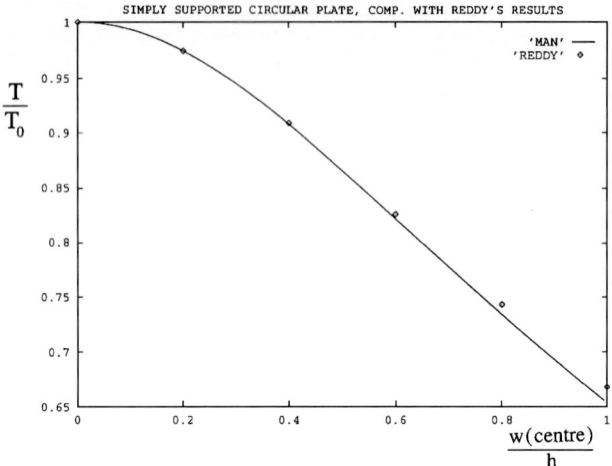

Figure 4. Period ratio $T/T_0 = \omega_0/\omega$ versus the displacement at the centre of a simply supported circular plate ($R/h = 100$, $\nu = 0.3$. R is the radius). Comparison with results of Reddy *et al.* [13] ◇ and Venkateswara *et al.* [12] ◇.

supported and clamped square plates. In Figures 4 and 5, we present a comparison with Reddy's results[12,13] in the case of circular plates. All these results are compared with published results. It can be seen from these figures that the agreement between the present results and the author's results is

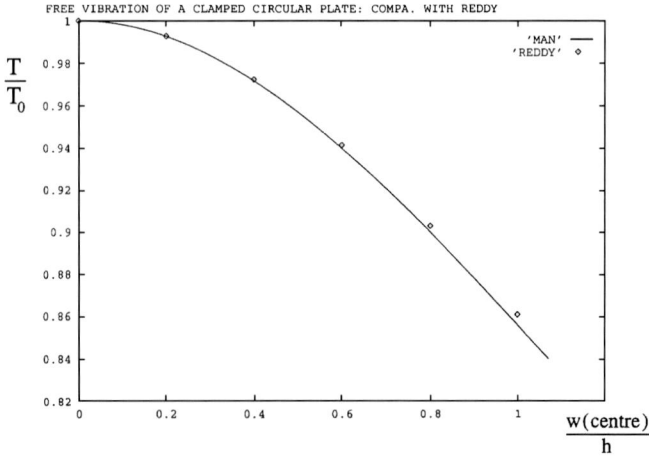

Figure 5. Period ratio T/T_0 versus the displacement at the centre of a clamped circular plate ($R/h = 100$, $\nu = 0.3$. R is the radius). Comparison with results of Reddy *et al.* [13] ◇ and Venkateswara *et al.* [12] ◇.

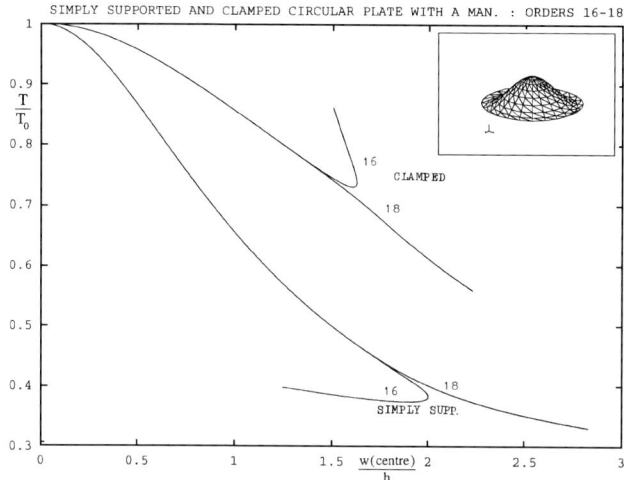

Figure 6. Frequency ratio versus the displacement at the centre of isotropic simply supported and clamped circular plate at orders 16 and 18 of the asymptotique series.

fairly good. In Figure 6, we present the ratio of period against the deflection for simply supported together with clamped circular plates until the radius of convergence. It is seen clearly that the period decreases with increasing the deflection and this tendency is most pronounced for simply supported plates. We note that the zone of convergence, which is given automatically by the series (48) is large in simply supported case than in clamped one. In Figures 7 and 8, we present the ratio $\dfrac{T_{NL}}{T_L}$ against the deflection $w\left(\dfrac{L}{4}, \dfrac{L}{4}\right)$ and the in-plane displacement $u_1\left(\dfrac{L}{4}, \dfrac{L}{4}\right), u_2\left(\dfrac{L}{4}, \dfrac{L}{4}\right)$ respectively at a quarter of the plate. (L is the side of the square plate). All these results are given in the zone of convergence. To study the nonlinear vibrations of plates with various shapes the algorithm needs only a file containing elements and nodes of the discretisation of the plate and imposed boundary conditions. For a test, we present in Figure 9 the backbone curve of an annular plate.

4.7. IMPROVEMENT OF THE SOLUTION BY USING PADE APPROXIMANTS

Until now, we are able to get a large part of the backbone curve for about the same computing time as one step of the modified Newton-Raphson algorithm. Although, this result is appreciable, one can be disappointed in

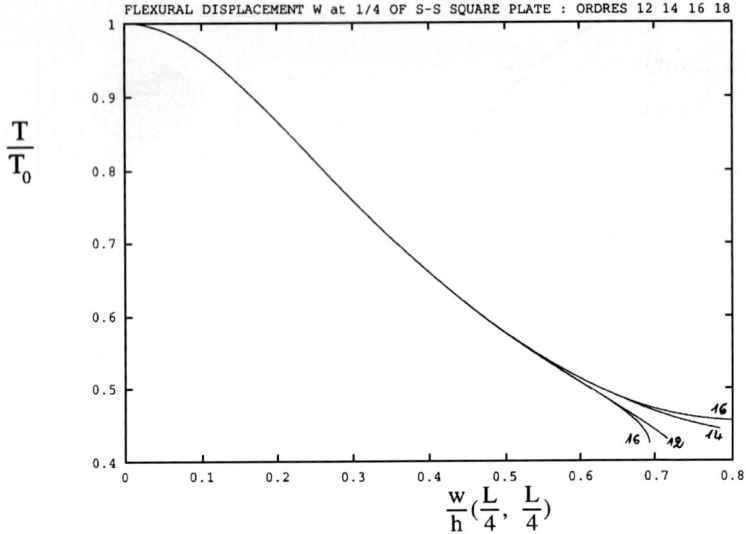

Figure 7. Frequency ratio versus the displacement $w(L/4,L/4)$ of a simply supported square plate.

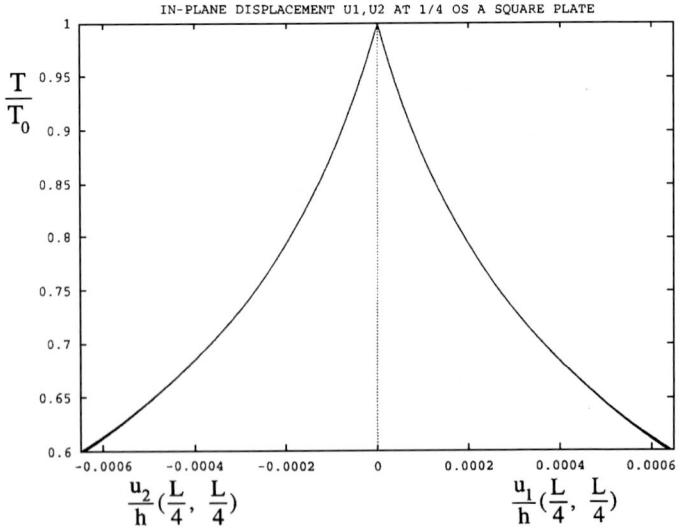

Figure 8. Frequency ratio versus the in-plane displacement $u_1(L/4,L/4)$, $u_2(L/4,L/4)$
of a simply supported square plate.

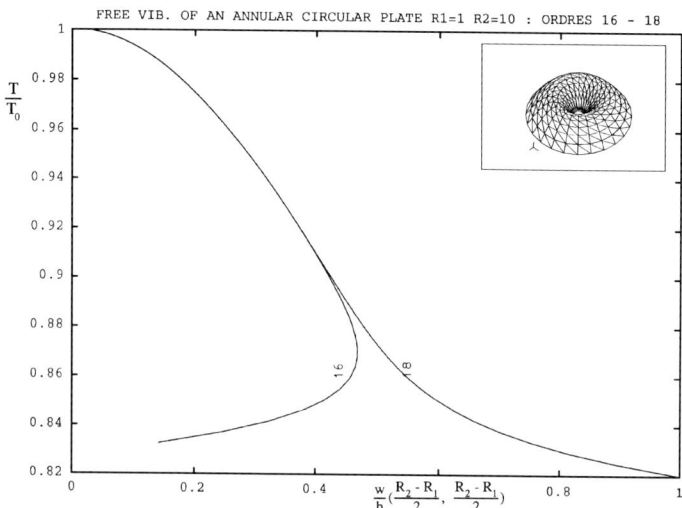

Figure 9. Frequency ratio versus the deflection $w[(R_2 - R_1)/2, (R_2 - R_1)/2]$ of an isotropic annular simply supported plate $R_1 = 1$ and $R_2 = 10$ by the A.N.M. at orders 16 and 18.

that the radius of convergence is not very large. As shown in the previous sections, this method permits us to compute a considerable number of the coefficients of the perturbed series. For improving the obtained solution we have to manipulate these coefficients. Fortunately, there are powerful methods for recovering an accurate approximation to the exact solution from a few terms of the perturbed series. These methods are crucial because they justify the study of perturbation theory; without them, perturbation theory would be effective just for local analysis. Therefore, for increasing the range of validity of the solution, several variants of the Asymptotic-Numerical Method had investigated in static problems of beams, plates and shells.[33,35,37]

The results we shall present here are done in static cases and give the post-buckling behaviour of plates and shells. Indeed, we do not have yet validated results in the dynamic case. In the static case the problem that we solved is written in the following sample form:

$$L.U + Q(U,U) = \lambda \, F$$

$$< L.U, \delta U > = \int_\Omega [N : \delta\gamma^1 + (\gamma^1 - C_m^{-1} : N) : \delta N + \delta\kappa : C_b : \kappa] \, d\Omega \qquad (49)$$

$$< Q(U,U), \delta U > = \int_\Omega [N : \delta\gamma^{nl} + \gamma^{nl} : \delta N] \, d\Omega$$

U is the mixed displacement and λ is the load parameter.

We assume that $(U_{(0)}, \lambda_{(0)})$ is a solution of equation (49) (for $\lambda = 0$, $U_{(0)} = 0$, $\lambda_{(0)} = 0$) and that in the vicinity of this point the branch can be represented by a power series with respect to a path parameter "a".

$$U_{(a)} = U_{(0)} + aU_{(1)} + a^2 U_{(2)} + a^3 U_{(3)} + \ldots \tag{50a}$$

$$\lambda_{(a)} = \lambda_{(0)} + a\lambda_{(1)} + a^2\lambda_{(2)} + a^3\lambda_{(3)} + \ldots \tag{50b}$$

Here, in static case, the computation procedure is the same as presented above. The power series (50) give the initial post-buckling behaviour of beams, plates or shells.[32–38]

In this section we shall show that the computed series (50) can be a-posteriori transformed into much more accurate approximation of the exact solution. As shown in numerical results of section V, we see clearly that the series (50) fail beyond the radius of convergence. An obvious reason to explain this failure could be that the n first vectors $V(p)$ form a too poor basis to represent the branch. An orthogonalisation of vectors $V(p)$ leading to an orthogonal basis seems to be more accurate. The coefficient of each new basis vector is a polynomial function. The basic idea for the improvement of the solution is to replace these polynomials by rational functions (ratio of two polynomials) called *Padé approximants*. The representation of a function by rational functions is in fact an old purpose that has been widely studied by Padé in his thesis (1892). An update representation of this subject can be found in the book of Backer and Graves-Morris.[42]

For the sake of simplicity, let us consider only the series of transverse displacement W truncated at order 5.

$$W = aW_{(1)} + a^3 W_{(3)} + a^5 W_{(5)} \tag{51}$$

The orthogonality condition (35) implies that $W(3)$ and $W(5)$ are both orthogonal to $W(1)$. But $W(5)$ is not a-priori orthogonal to $W(3)$ and we separate it into a part which is collinear to $W(3)$ and a part which is orthogonal.

$$W_{(5)} = \alpha_{35} W_{(3)} + W^{\perp}_{(5)}$$
$$< W^{\perp}_{(5)}, W_{(3)} > = 0 \tag{52}$$

Hence (51) can be written as

$$W = aW_{(1)} + (a^3 + \alpha_{35}\, a^5)W_{(3)} + a^5 W^{\perp}_{(5)} \tag{53}$$

The basic idea for the improvement of the solution is to replace the polynomial $(a^3 + \alpha_{35}\, a^5)$ by the rational function $\dfrac{a^3}{1 - \alpha_{35} a^2}$.

These representations are asymptotically equivalent for small "a". But the fraction grows as "a" instead of "a^5" what looks more like the exact solution. Hence, the new representation with a rational fraction is

$$W = a\, W_{(1)} + \frac{a^3}{1 - \alpha_{35} a^2}\, W_{(3)} \tag{54}$$

It is asymptotically equivalent to (53) at order 3 but it fits the exact solution much better than (53) can do.

The generalisation of this representation to the complete series (50a) can be easily done. First, we built up an orthogonal basis $U_{(p)}^{\perp}$ from the vectors $U_{(p)}$ by a classical Gram-shmidt orthogonalisation procedure.

$$U(p) = \sum_{k=1}^{p} \alpha_p^k \qquad \alpha_p^k = \frac{<U(p), U^{\perp}(k)>}{<U^{\perp}(k), U^{\perp}(k)>} U_{(k)}^{\perp} \qquad \text{for } 1 < k < p$$

$$\alpha_p^p = 1 \quad \text{and} \quad \alpha_p^1 = 0 \tag{55}$$

and we rewrite (50a) with the new orthogonal basis.

$$U(a) = \sum_{j=1}^{p} a^j\, f_j(a)\, U_{(j)}^{\perp} \qquad f_j(a) = \sum_{j=1}^{p} \alpha_i^j\, a^{i-j} \tag{56}$$

Second, we replace the scalar series $f_j(a)$ by their Padé approximant, that is the rational function:

$$P_j[L_j / M_j](a) = \left(\frac{b_0^j + b_1^j a + \ldots + b_{Lj}^j a^{Lj}}{1 + c_1^j a + \ldots + c_{Mj}^j a^{Mj}} \right) \tag{57}$$

which has the same Mac-Laurin expansion as (56) at the order $Mj + Lj + 1$. The $Mj + Lj + 1$ coefficients b_i^j and c_i^j for each j can be computed by solving linear systems.[42] Thus, we have defined a rational function representation of the solution $U(a)$

$$U(a) = \sum_{j=1}^{\infty} a^j\, P[Lj / Mj]\, U_{(j)}^{\perp} \tag{58}$$

The representation of the load parameter can be easily obtained as follows:

$$\lambda(a) - \lambda_{(0)} = -\frac{< Q(U(a),U(a));\ U(1) >}{< L'.U(a),\ U(1) >} \tag{59}$$

Finally, the solution of the problem (49) is given by the rational representations (58–59). Let us recall that the manipulation of the vectors $U_{(p)}^{\perp}$ and the Padé approximants $Pj[Lj/Mj]$ requires a few computation time. Only the orders Lj and Mj of each Padé approximant (57) have to be chosen.[35,37] In

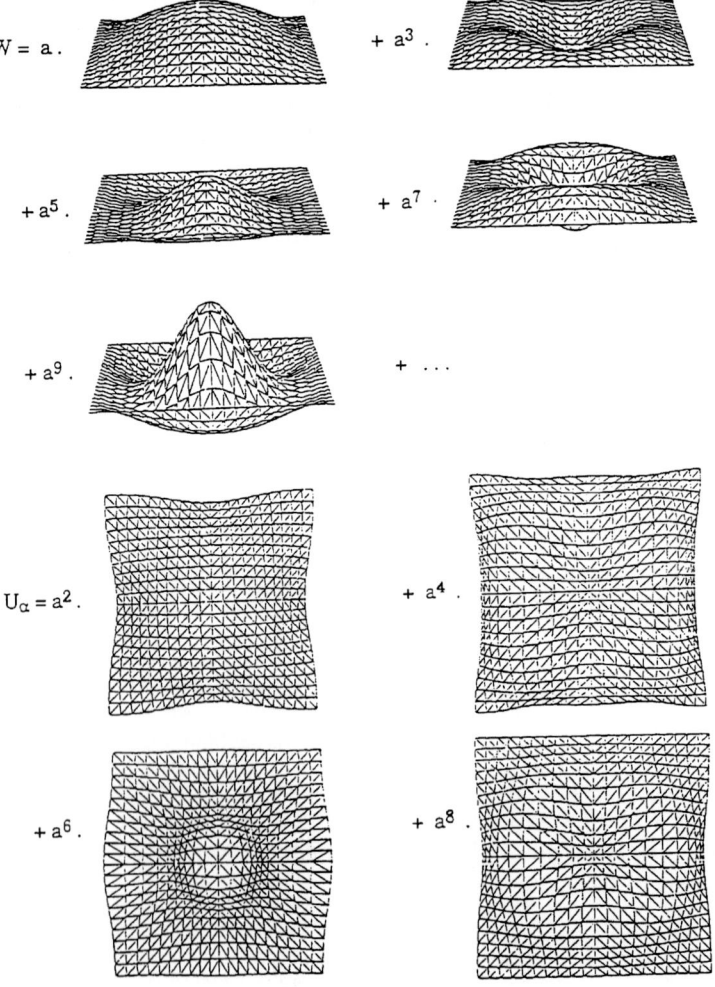

Figure 10.

order to limit the number of poles of these Padé approximants and to increase the robustness of this algorithm a new way to build up these approximants has been presented.[47]

Numerical Results

We present here a test of this procedure on the academic problem of simply supported square plate loaded by a uniform uni-axial compression. In Figure 10, we present the deflection $W(p)$ for the odd problems and the in-plane displacement $U_\alpha(p)$ for the even problems. The visualisation of these vectors shows clearly that these vectors are not independent. So the orthogonalisation of them leads to an orthogonal basis which permits to use the effective contribution of each vector. In Figure 11, we plot the ratio W/h versus the ratio λ/λ_c of the series (50) truncated at orders 2–4–6–8. W is the deflection at the centre of the plate, h is the thickness of the plate, λ_c is the buckling load and λ is the load parameter. The solutions are compared with the exact solutions obtained by the Newton-Raphson algorithm. For the load displacement problem of plate the radius of convergence is about $\lambda/\lambda_c = 1.4$ and $w/h = 1.6$ which is as the same order than in the previous studied nonlinear vibration problems. Already now, we are able to get a large part of the bifurcating branch for about the same computing time as one step of the modified Newton-Raphson algorithm.

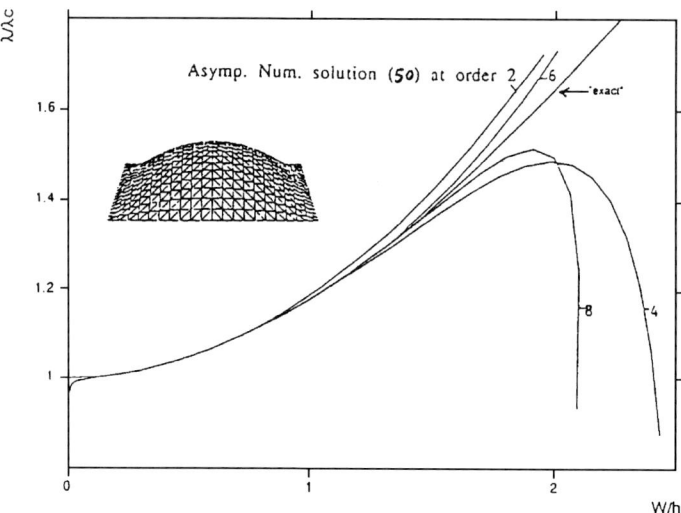

Figure 11. Load-displacement curve for the plate: W/h at the centre versus the load ratio λ/λ_c. Comparison between the "exact" solution and different truncatures 2–4–6–8 of the asymptotic-numerical solution.

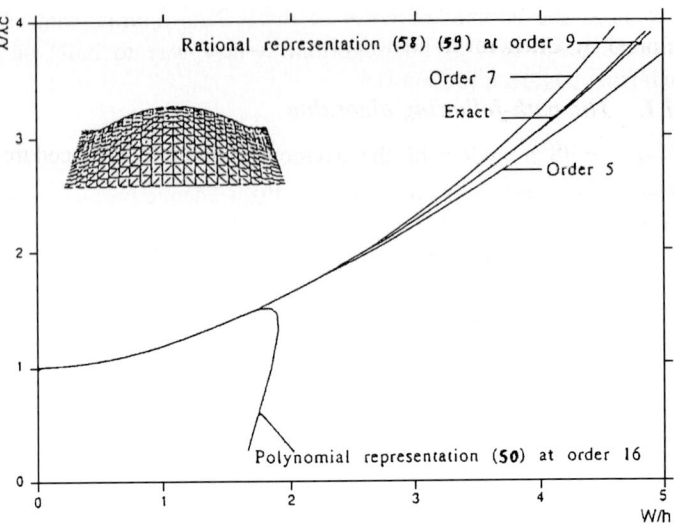

Figure 12.　Improvement of the solution using Padé approximants. By this way we are able to compute the solution up to $\lambda = 4\lambda c$ and $w = 5h$.

For improving this zone of validity of the solution we use the Padé approximants. This procedure requires very few computing time. In Figure 12, we can see that, by this way we are able to compute the bifurcating branch of the plate up to $\lambda = 4\lambda_c$ using 19 terms of the series (50) and only 9 terms after orthogonalisation. Then this procedure improves considerably the solution.

4.8.　CONTINUATION PROCEDURE

The path-following techniques are well known with iterative-incremental procedure like arc length control. The principle is to follow the nonlinear solution branch in a stepwise manner via a succession of iterations to achieve equilibrium and incrementations of a proper parametrisation of the path. The solution is given numerically point by point. With the Asymptotic-Numerical Method, which is a perturbation method, the solution is given analytically as a function of the perturbed parameter. It has been shown that the validity of these solutions is limited by a radius of convergence which is not necessary small. Taking a starting point in the zone of validity of the solution, we can reapply the A.N.M. and go far in the solution path. Although, the continued solution has a radius of convergence, the application of the A.N.M. iteratively permits us to determine a complex nonlinear branch by a succession of local asymptotic expansions. This is the idea developed by Cochelin in static case.[38] and will be presented here.

1.8.1.1. *The path-following algorithm*

We begin with a review of the asymptotic-numerical procedure which provides an analytical representation of nonlinear branch in a neighbourhood of a starting point $(U_{(0)}, \lambda_{(0)})$. By introducing equation (50) into equation (49) and equating like powers of "a" we obtain the following set of linear mixed problems:

$$
\begin{aligned}
\text{order 1:} \quad & L_t.U_{(1)} = \lambda_{(1)} F \\
\text{order 2:} \quad & L_t.U_{(2)} = \lambda_{(2)} F - Q(U_{(1)}, U_{(1)}) \\
& \vdots \\
\text{order } p: \quad & L_t.U_{(p)} = \lambda_{(p)} F - \sum_{r=1}^{p-1} Q(U_{(r)}, U_{(p-r)})
\end{aligned}
\tag{60}
$$

where the tangent operator L_t is defined as $L_t(.) = L(.) + 2 Q(U_{(0)}, ..)$. The first equation corresponds to the linearisation of equation (49) at the starting point $(U_{(0)}, \lambda_{(0)})$. i.e. the vector $U_{(1)}$ and the coefficient $\lambda_{(1)}$ correspond to the tangent of the branch at the starting point. Notice that at each order p both $U_{(p)}$ and $\lambda_{(p)}$ are unknown and there is one superfluous unknown in each of these linear problems. So, we must add a solvability equation. Therefore, it is better to consider a measure that includes the entire displacement vector and also the load parameter, i.e. an arc length measure.[44-46]

Following this idea, we shall identify the path parameter "a" as the projection of the displacement increment $(U_{(a)} - U_{(0)})$ and the load increment $(\lambda_{(a)} - \lambda_{(0)})$ on the tangent vector $(U_{(1)}, \lambda_{(1)})$[38]

$$
a = \frac{1}{s^2} \{< U - U_{(0)}, U_{(1)} > + (\lambda - \lambda_{(0)})\lambda_{(1)}\}
\tag{61}
$$

Here $<.,.>$ is the Euclidean scalar product and s is a scaling parameter which corresponds to the length of the tangent vector $(U_{(1)}, \lambda_{(1)})$. Introducing the series (50) into equation (56) and equating like powers of "a" we obtain the following set of single equations

$$
\begin{aligned}
\text{order 1:} \quad & < U_{(1)}, U_{(1)} > + \lambda_{(1)}\lambda_{(1)} = s^2 \\
\text{order 2:} \quad & < U_{(1)}, U_{(2)} > + \lambda_{(1)}\lambda_{(2)} = 0 \\
\text{order 3:} \quad & < U_{(1)}, U_{(3)} > + \lambda_{(1)}\lambda_{(3)} = 0 \\
& \vdots \\
\text{order } p: \quad & < U_{(1)}, U_{(p)} > + \lambda_{(1)}\lambda_{(p)} = 0
\end{aligned}
\tag{62}
$$

Finally, all vectors $U_{(p)}$ and coefficients $\lambda_{(p)}$ of the series (50) can be determined by successively solving the systems of equations (60) and (62) at each order.[38]

In order to have an efficient algorithm, the analysis of the range of validity and definition of a new starting point should be automatic, i.e. we have to automatically determine the values a_m of the path parameter over which the polynomial solution will not satisfy a given accuracy. We recall that the polynomial solutions are very similar inside the radius of convergence of the series, but they separate rapidly when this radius is reached. So, a simple criterion is to require that the difference between two consecutive order solutions remains small.

$$\frac{\left\| U_{\text{order } N} - U_{\text{order } N-1} \right\|}{\left\| U_{\text{order } N} - U_0 \right\|} = \frac{\left\| a^N U_N \right\|}{\left\| a\, U_1 + a^2 U_2 + \ldots + a^N U_N \right\|} \le \varepsilon \tag{63}$$

Here, ε is a small number. By approximating the denominator as $\| aU_{(1)} \|$ we obtain a simple criterion in displacement.

$$a_m = \varepsilon \left(\frac{\| U_1 \|}{\| U_N \|} \right)^{\frac{1}{N-1}} \tag{64}$$

Note that, this simple criterion gives a good order of magnitude of the validity of the solution, whereas it requires almost no computing time. However, a more secure way of controlling the quality of the asymptotic solution consists of computing residual vectors. Then a_m can be determined with some interpolation techniques. Anyway, the simple criterion given above is very helpful in defining the range of interest of "a" so that the computation of the residual vectors can be limited to very few points.

Finally, we show that this method can be applied iteratively in a step by step manner. Because the local analytical representation of the branch within each step the present path following technique has some important advantages compared to classical predictor-corrector schemes. Particularly, the algorithm is very robust and completely automatic.

Numerical Results

To illustrate the effectiveness of this path-following technique, numerical results of a problem of shell instability including snap-through and snap-back phenomena will be presented. The test considered here is the well known problem of a cylindrical shallow shell loaded by a single force which is a classical benchmark in nonlinear shell analysis. We used a 10×10 mesh

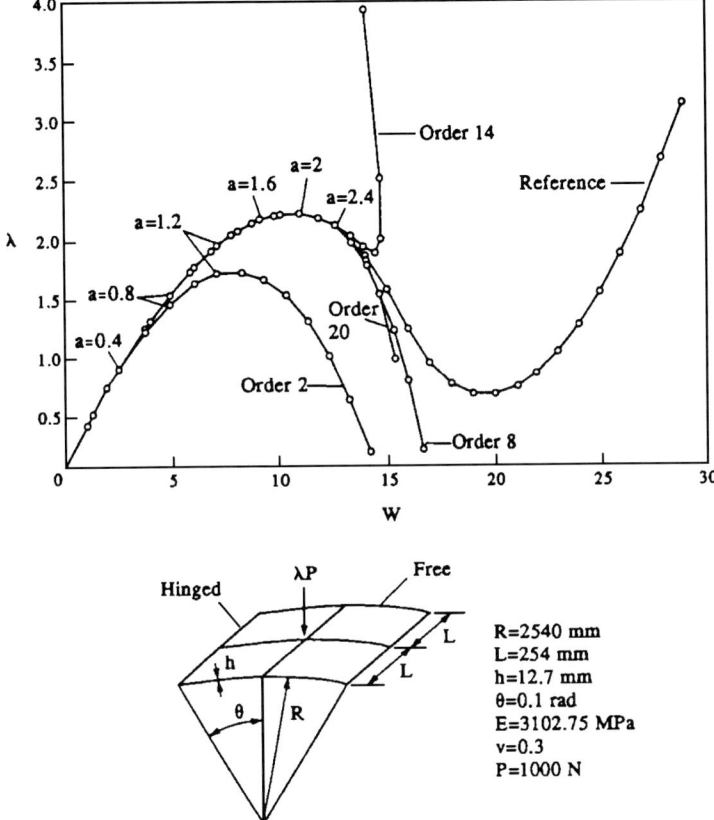

Figure 13. Cylindrical shell under point load: comparison between the polynomial approximations at order 2, 8, 14 and 20 and the exact solution. The values of the path paramter 'a' have been reported all along these curves.

for a quarter of shell and 200 triangular DKT elements. Figure 13 shows the load-deflection curves for the asymptotic solutions at orders 2,8,14,20 and for the reference solution that has been obtained by Newton-Raphson method. The values of the path parameter "a" have been reported in the figure. We can see that the asymptotic numerical solutions become unacceptable when "a" is greater than about 2.4 or 2.6. We can easily define a new starting point by computing $U(a = 2)$ and $\lambda(a = 2)$. By reapplying the asymptotic procedure from this new point we shall determine a further part of the branch. Practically, the starting point $U(a_m)$ and $\lambda(a_m)$ is automatically determined by equation (64). The only parameters that the user has to provide are the

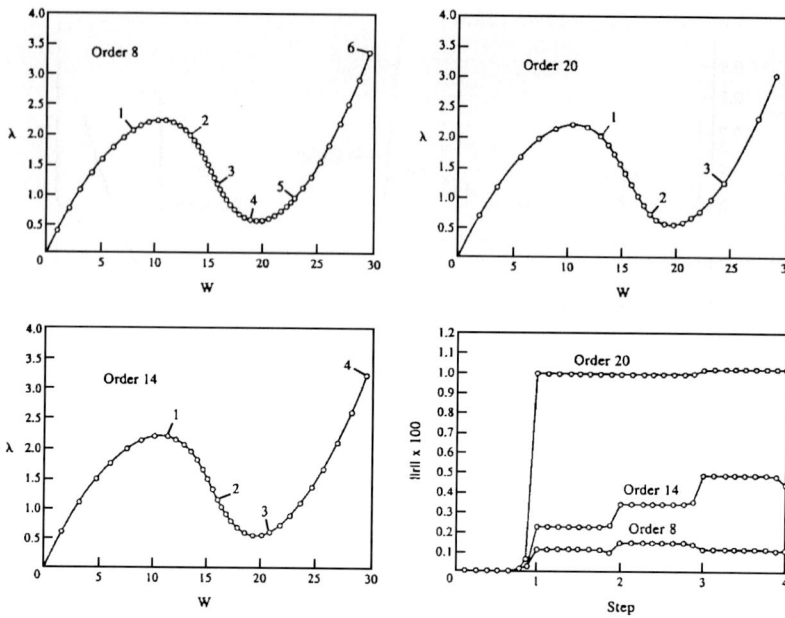

Figure 14. Cylindrical shell under point load: results of the continuation for various order of truncature N and a fixed $\epsilon = 10^{-4}$. All these curves have been plotted by using eight points for each step. The last point of each step is indicated by a number; (d) shows the evolution of the norm of the residual vector along the steps.

order N, the precision parameter ε and the number of steps. Figure 14 show the load displacement curves for various orders of truncature N with a fixed criterion $\varepsilon = 10^{-4}$. The norm of the residual vector corresponding to these curves is also given in Figure 14d. At order $N = 8$, it requires six steps to describe the complete curve up to $\lambda = 3$. Only four steps are needed at order 14 and let us say three and half at order 20. Figure 15 shows the results for the same shell problem but with a smaller thickness. At order $N = 20$, with $\varepsilon = 10^{-4}$, the full curve is obtained in 11 steps, which means with 11 stiffness matrix decompositions. There exists a great number of references where this test has been solved by a Newton-Raphson method. Another test often used by the authors is the simply supported cylindrical shell loaded by a uniform axial compression. The response is shown in Figure 16, which represents the axial displacement versus the load parameter. We can see clearly the automatic adaptation of the steps along the branch.

These examples demonstrate the applicability of the Asymptotic-Numerical Method to a wide range of nonlinear engineering problems.

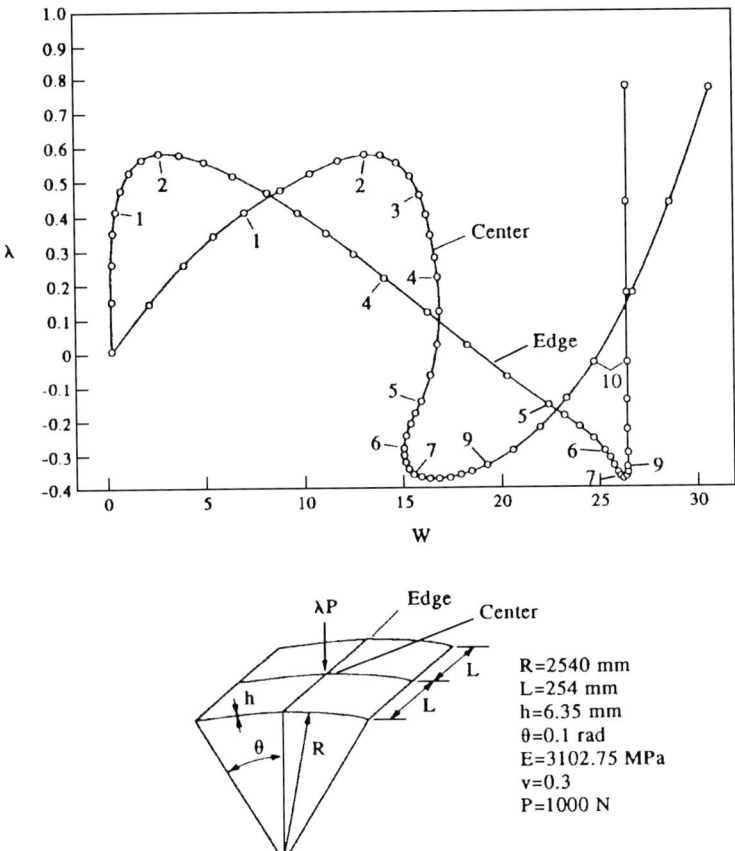

Figure 15. Cylindrical shell under point load: same problem as in Figure 5 but with a smaller thickness for the shell. The two curves correspond to the displacement at the centre and at the edge of the shell. At order 20, with $\epsilon = 10^{-4}$, the branch is obtained within 11 steps (four points per step). The norm of the residual vector at the end of the branch is about 10^{-3}.

4.9. CONCLUSION

The applicability of the Asymptotic-Numerical Method to study the nonlinear vibrations of plates with various geometry has been demonstrated. Some sophisticated method can be easily added to improve the range of validity of the obtained solution especially for more complex nonlinear behaviours. Some examples in static cases are presented to demonstrate the applicability

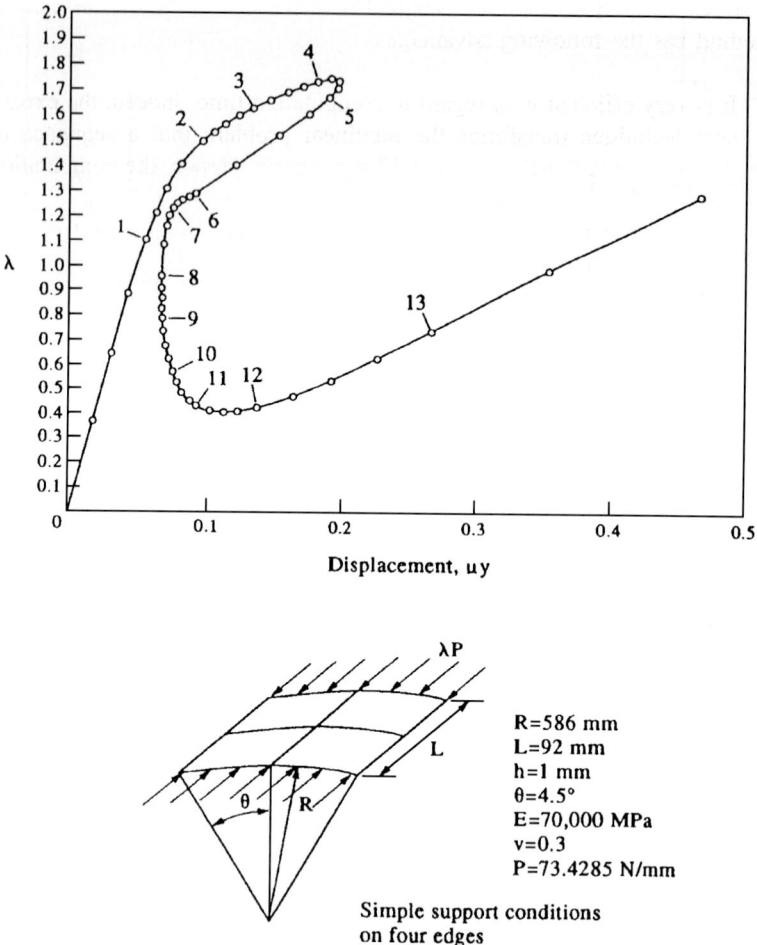

Figure 16. Cylindrical shell under compression: order 20, with $\epsilon = 10^{-4}$, the branch is obtained within 14 steps. The residual at the end is about 3×10^{-3}.

of this method to a wide range of nonlinear engineering problems. The principle of this method is to compute numerically two polynomials series that give the displacement and frequency as a function of a perturbation parameter. By using a mixed formulation, the governing equation has been written with a quadratic nonlinearity. Hence, the expansion procedure is rather simple and it can be easily implemented in an existing F.E. software. It permits one to compute a very large number of terms of the series.

In comparison with a more classical step by step procedure, the present method has the following advantages:

- It is very efficient with regard to computation time. Indeed, the expansion technique transforms the nonlinear problem into a sequence of linear problems with a single stiffness matrix. Hence, the computation time is of the same order as for a single step of the modified Newton-Raphson algorithm.
- The solution branch is known continuously and not only on some points.
- The computation of the series is fully automatic. The only parameter that has to be chosen is the order of truncature.
- There is no need to decide *a priori* the step length as in step by step procedure. The range of validity of the step length is given by the method itself and it is not chosen *a priori* by the user.
- In comparison with classical perturbation methods a great number of terms of the perturbed series can be easily and automatically computed. Hence, some procedures for improving the range of validity of the solution, such as Padé approximants, can be easily added without more computation time.
- A path following technique via this asymptotic numerical method is successfully used and permits to study strongly nonlinear problems.

In summary, this method is efficient and reliable. It has been illustrated here on nonlinear vibrations of plates and on some static post-buckling behaviours of plates and shells. Other related topics such as forced nonlinear vibrations, internal resonance and nonlinear damped vibrations of thin elastic structures will be investigated.

References

1. Yamaki, N., 1961, "Influence of large amplitudes on flexural vibrations of elastic plates". *ZAMM*, **41**, 501–510.
2. Wah, T. and M. Asce, 1963, "Vibration of circular plates at large amplitudes". *J. Engin. Mech. Division*, 1–15.
3. Eisley, J.G., 1964, "Nonlinear vibration of beams and rectangular plates". *ZAMP*, **15**, 167–175.
4. Srinivasan, A.V., 1966, "Nonlinear vibrations of beams and plates". *Int. J. Nonlin. Mech.*, **1**, 179–191.
5. Tamura, H., T. Okabe and A. Sueoka, 1993, "Exact solutions of the free vibration of a system with asymmetrical single-term cubic spring". *JSME*, **36**(1), 26–34.
6. Bert, C.W., 1982, "Research on dynamics of composite and sandwich plates 1979–81". *Shock Vib. Dig.*, **14**(10), 17–34.
7. Sathyamoorthy, M., 1983, "Nonlinear vibrations of plates, a review". *Shock Vib. Dig.*, **15**, 3–16.
8. Chia, C.Y., 1980, "Nonlinear analysis of plates", Mc-Graw Hill, New York.

9. Nageswara Rao, B. and S.R.R. Pillai, 1992, "Nonlinear vibrations of a simply supported rectangular antisymmetric cross-play plate with immovable edges". *J. Sound and Vibration*, **152(3)**, 568–572.

10. Nageswara Rao, B. and S.R.R. Pillai, 1992, "Large amplitude free vibrations of laminated anisotropic thin plates based on harmonic balance method". *J. Sound and Vibration*, **154(1)**, 173–177.

11. Mei, C., 1973, "Finite element displacement method for large amplitude free flexural vibrations of beams and plates". *Computer & Structures*, **3**, 163–174.

12. Venkateswara Rao, G., K. Kanaka Raju and I.S. Raju, 1976, "Finite element formulation for the large amplitude free vibrations of beams and orthotropic circular plates". *Computer & Structures*, **6**, 169–172.

13. Reddy, J.N. and C.L. Huang, 1981, "Large amplitude free vibrations of annular plates of varying thickness". *J. Sound and Vibration*, **79(3)**, 387–396.

14. Reddy, J.N., C.L. Huang and I.R. Singh, 1981, "Large deflections and large amplitude vibrations of axisymmetric circular plates". *Int. J. Numer. Meth. Engng.*, **17**, 527–541.

15. Mei, C. and Kamolphan Decha-Umphai, 1985, "A finite element method for nonlinear forced vibrations of rectangular plates". *AIAA J.*, **23(7)**, 1104–1110.

16. Kant, T. and J.R. Kommineni, 1994, "Large amplitude free vibration analysis of cross-ply composite and sandwich laminates with a refined theory and C° finite elements". *Computer & Structures*, **50**, 123–134.

17. Carter Wellford L.J.R., G.M. Dib and W. Mindle, 1980, "Free and steady state vibration of nonlinear structures using finite element nonlinear eigenvalue technique". *Earth. Engi. Stru. Dyna.*, **8**, 97–115.

18. Noor, A.K., 1981, "Recent advances in reduction methods for nonlinear problems". *Computer & Structures*, **13**, 31–44.

19. Noor, A.K., C.M. Andersen, J.M. Peters, 1993, "Reduced basis technique for nonlinear vibration analysis of composite panels". *Comput. Meth. Appl. Mech. Eng.*, **103**, 175–186.

20. Lewandowski, R., 1987, "Application of the Ritz method to the analysis of nonlinear free vibrations of beams". *J. Sound and Vibration*, **114(1)**, 91–101.

21. White, R.G. and C.E. Teh, 1981, "Dynamic behaviour of isotropic plates under combined acoustic excitation and static in-plane compression". *J. Sound and Vibration*, **75**, 527–547.

22. Benamar, R., M.M.K. Bennouna and R.G. White, 1993, "The effects of large vibration amplitudes on the mode shapes and natural frequencies of thin elastic structures. Part II: fully clamped rectangular plates". *J. Sound and Vibration*, **164**, 399–424. Part III, 1994, Fully clamped rectangular isotropic plates measurements of the mode shape amplitude dependence and the spatial distribution of harmonic distortion". *J. Sound and Vibration*, **175(3)**, 377–395.

23. Benamar, R., 1990, Ph. D. Thesis, University of Southampton "Nonlinear dynamic behaviour of fully clamped beams and rectangular isotripic and laminated plates".

24. Benamar, R. and R.G. White "A new semi analytical approach to the nonlinear dynamic forced response problem of fully clamped beams and rectangular plates at large vibration amplitudes. Part I: General theory and application to the one mode case" To appear.

25. Azrar, L. and R. Benamar, 1995, "Etude des vibrations non linéaires forcées des poutes par une méthode semi-analytique" 5ème Colloques Maghrébin sur les Modèles Numériques de l'Ingénieur, E.M.I., Rabat, Maroc, 594–599.

26. Nayfeh, A.H., 1973, "Perturbation methods", John Wiley and Sons N.Y.

27. Sridhar, S., D.T. Mook and A.H. Nayfeh, 1975, "Nonlinear resonances in the forced responses of plates. Part I: Symmetric resonances of circular plates". *J. Sound and Vibration*, **41(3)**, 359–373. Part II, 1978, asymmetric responses of circular plates". *J. Sound and Vibration*, **59(2)**, 159–170.

28. Nayfeh, A.H., S.A. Nayfeh, 1994, "On nonlinear modes of continuous systems". *J. Vibration and Acoustics*, **116**, 129–136.

29. Cheung, Y.K. and S.L. Lau, 1982, "Incremental time-space finite strip method for nonlinear structural vibrations". *Earth. Engin. and Structural Dynamics*, **10**, 239–253.
30. Leung, A.Y.T. and S.G. Mao, 1995, "Symplectic integration of an accurate beam finite element in nonlinear vibration". *Computer & Structures*, **54**, 1135–1147.
31. Aprile, A., A. Benedetti and T. Trombetti, 1994, "On nonlinear dynamic analysis in the frequency domain: Algorithms and applications". *Earth. Eng. and Strut. Dynamics*, **23**, 363–388.
32. Damil, N. and M. Potier-Ferry, 1990, "A new method to compute perturbed bifurcation: Application to the buckling of imperfect elastic structures". *Int. J. Engng. Sci.*, **28**, 943–957.
33. Azrar, L., B. Cochelin, N. Damil, M. Potier-Ferry, 1992, "An Asymptotic Numerical Method to compute bifurcating branches" In P. Ladevèze and O.C. Zienkiewicz, editors, *New Advances in computational structural mechanics*, pp. 117–131. Elsevier.
34. Azrar, L. and M. Potier-Ferry, 1992, "Post-buckling of imperfect structures by an Asymptotic-Numerical Method" Numerical Methods in Ch Hirsch. *et al.*, editors, *Engineering*, pp. 607–617. Elsevier.
35. Azrar, L., 1993, "Etude du comportement post-critique des coques cylindriques par une Méthode Asymptotique Numérique ", Thèse de doctorat de l'université de Metz.
36. Azrar, L., B. Cochelin, N. Damil, M. Potier-Ferry, 1993, "An Asymptotic Numerical Method to compute the post-buckling behaviour of elastic plates and shells". *Int. J. Numer. Meth. Engng.*, **36**, 1251–1277.
37. Cochelin, B., N. Damil, M. Potier-Ferry, 1994, "Asymptotic-Numerical Method and Padé approximants for nonlinear elastic structures". *Int. J. Numer. Meth. Engng.*, **37**, 1187–1213.
38. Cochelin, B., 1994, "A path-following technique via an Asymptotic-Numerical Method". *Computer & Structures*, **53**(5), 1181–1192.
39. Zienkiewicz, O.C., 1977, "The finite element method "3rd ed, McGraw-Hill, London.
40. Batoz, J.L., K.J. Bathe and L.W. Ho, 1980, "A study of three node triangular plate bending elements". *Int. J. Numer. Meth. Engng.*, **15**, 1771–1812.
41. Lau, S.L., Y.K. Cheung, S.Y. Wu, 1984, "Nonlinear vibration of thin elastic plates. Part I: Generalised incremental Hamilton's principle and finite element formulation". *J. Appl. Mech.*, **51**, 837–844.
42. Baker, G.A. and P. Graves-Morris, 1981, "Padé approximants, Part I: basic theory". Encyclopedia of Mathematics and its applications, Vol. 13, Addison-Wesley.
43. Van Dyke, M., 1984, "Computer-extended series". *Ann. Rev. Fluid Mech.*, **16**, 287–309.
44. Crisfield, M.A., 1983, "An arc-length method including line searches and accelerations". *Int. J. Numer. Meth. Engng.*, **19**, 1269–1289.
45. Riks, E., 1984, "Some computational aspects of the stability analysis of nonlinear structures". *Comput. Meth. Appl. Mech. Engng*, **47**, 219–259.
46. Carrera, E., 1994, "A study on arc-length-type methods and their operation failures illustrated by a simple model". *Computer & Structures*, **50**(2), 217–229.
47. Najah, A., B. Cochelin, N. Damil, M. Potier-Ferry "A critical review of asymptotic numerical methods". To appear.

5 NONLINEAR ANALYSIS OF RECTANGULAR LAMINATED PLATES UNDER END SHORTENING USING THE FINITE STRIP METHOD

S.S.E. LAM[1] and D.J. DAWE[2]

[1] *Department of Civil and Structural Engineering,*
Hong Kong Polytechnic University, Hunghom, Kowloon, Hong Kong
[2] *School of Civil Engineering, The University of Birmingham,*
Edgbaston, Birmingham B15 2TT, United Kingdom

5.1. INTRODUCTION

Laminated plates are increasingly used as structural components in various branches of engineering because of their high specific strength and specific stiffness. Examples of using laminated plates vary from the inexpensive plywood and cardboard boxes to the expensive highly sophisticated structural components in aerospace and marine structures.[1] These plates are made of plies or laminae each consisting of parallel fibres embedded in a matrix material and are bonded together by suitable adhesive. Considerable variation and complication in plate configuration may arise by tailoring the layup arrangement of the laminae. Such flexibility increases the complexity in predicting the structural behaviour of the laminated plates under loading, as compared to situations prevailing with homogeneous isotropic materials.

Flat laminates are frequently used as primary structural components in situations where they are subjected to in-plane loading. The demand to optimise the weight of such structural components leads to a need to account for the nonlinear response of the laminates. However, available theoretical

143

work related to the prediction of the post-buckling strength of metallic panels are not directly applicable to laminated plates having orthotropic or anisotropic properties. There is a need to develop efficient analytical tool to predict accurately the response of the laminates and in particular the buckling and post-buckling responses. No attempt is made here to review the full range of work related to laminated plate buckling since recent review articles are available by Leissa[2] and Kapania and Raciti.[3]

Subject to the arrangement of the laminae, laminated plates may have relatively low through-thickness shear rigidity such that the use of the classical plate theory (CPT) may be invalid. The first-order shear deformation plate theory (SDPT) relaxes the normalcy condition assumed in the CPT by allowing the initial straight normals to the middle surface to remain straight during deformation, but are not constrained to remain normal to the deformed middle surface.[4–5] In the field of linear analysis the finite strip method has been developed extensively by Dawe and co-workers for the analysis of complicated laminated plate structures based on the use of both the SDPT and CPT.[6–9] In geometrically nonlinear analyses, Graves-Smith and co-workers[10–12] and Hancock[13] and Bradford and Hancock[14] have used the finite strip method in the context of CPT to predict the response of homogeneous single plates and plate structures, usually for post-buckling behaviour. Dawe and co-workers[15–20] have extended the finite strip method in geometrically nonlinear analysis of laminates based on the use of the SDPT and CPT. More recently, Dawe and co-workers[21] have studied the nonlinear response of laminates subjected to progressive end shortening whilst initial imperfection and/or applied pressure loading are also present.

This paper is concerned with the development of the SDPT and CPT finite strips on the prediction of the elastic post-buckling response of flat rectangular laminates with general material properties. The laminates are subjected to progressive uniform end shortening with the loaded ends of the laminate simply supported out-of-plane. A range of applications is presented, involving isotropic plates and orthotropic and anisotropic laminates with balanced and unbalanced layups.

5.2. FINITE STRIP METHOD

The finite strip method, introduced by Cheung,[22,23] is now well established for the analysis of single plates and prismatic structures. It is a potential energy (or virtual work) approach lying between the traditional single-field Rayleigh-Ritz method and the finite element method. Figure 1 shows a finite strip with x and y axes in the plane of the strip and z axis in the thickness direction. The displacement field used for the displacement components, say w, is in a form similar to the use of separation of variables in solving partial

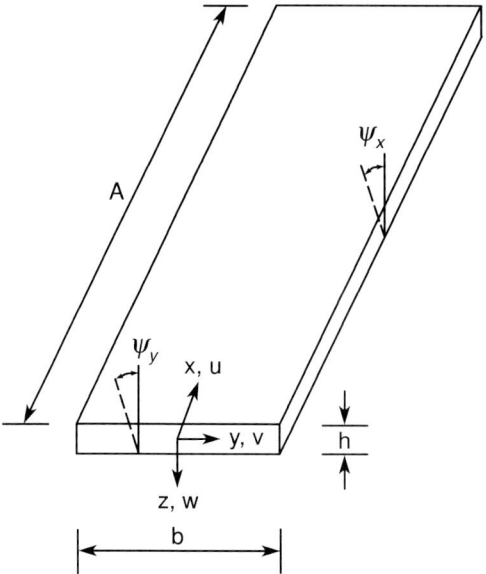

Figure 1. A typical finite strip.

differential equations.

$$w(x,y) = \sum_i W_i(x)g_i(y) \tag{2.1}$$

The variation of the displacement w in x and y directions is separated into two independent functions $W_i(x)$ and $g_i(y)$. $g_i(y)$ are expressed as piecewise polynomial functions similar to the interpolation functions used in one-dimensional finite elements. $W_i(x)$ are longitudinal series terms satisfying the kinematic conditions at the strip ends which can be taken as the characteristic solution of the beam vibration problem

$$\frac{d^4 W(x)}{dx^4} - \left(\frac{\mu}{A}\right)^4 W(x) = 0 \tag{2.2}$$

where μ is a parameter. The general solution of equation (2.2) is

$$W(x) = C_1 \sin\left(\frac{\mu x}{A}\right) + C_2 \cos\left(\frac{\mu x}{A}\right) + C_3 \sinh\left(\frac{\mu x}{A}\right) + C_4 \cosh\left(\frac{\mu x}{A}\right) \tag{2.3}$$

where the coefficients C_1 to C_4 are determined by consideration of the kinematic conditions.

This study is concerned with the finite strip analysis of prismatic plate structures with simply supported boundaries at the strip ends,

$$W(x) = 0 \quad \text{and} \quad \frac{d^2 W(x)}{dx^2} = 0 \quad \text{at} \quad x = 0, A \tag{2.4}$$

Substitution of equations (2.4) into (2.3), yields

$$W(x) = \sum_i \sin\left(\frac{i\pi x}{A}\right) \Rightarrow W_i(x) = \sin\left(\frac{i\pi x}{A}\right) \tag{2.5}$$

$W_i(x)$ are the mode shapes characterised by the orthogonal properties,

$$\int_0^A W_i(x) \, W_j(x) \, dx = 0, \quad \text{if} \quad i \neq j \tag{2.6}$$

Similar expressions can be found for other displacement components and trigonometric series in the form of a sine series or a cosine series (of course, subject to the satisfaction of the kinematic conditions at the strip ends) can be used to express the longitudinal series terms for the finite strips used in this study.

In linear analysis the orthogonal properties, equation (2.5), uncouple the finite strip stiffness for different longitudinal series terms and the resulting structural stiffness matrix has very narrow bandwidth. Such advantage disappears in nonlinear finite strip formulation and coupling between different longitudinal series terms in the finite strip occurs due to the presence of cubic and quartic terms in the strain energies, equations (3.8) or (4.2). For plates where post-buckling behaviour is important, it has already been well demonstrated that the finite strip method has the advantage of relative computational economy as compared with the finite element method in attending similar accuracy.[19] A significant part of the reason for this is that in the finite element method the bandwidth of structural stiffness matrix is strongly dependent on the mesh size, whereas in the finite strip method this is not so.

5.3. NONLINEAR FORMULATION BASED ON SDPT

Figure 1 shows a typical finite strip which forms part of a rectangular plate of length A. The response of the finite strip is assumed to be elastic. Implementing a first-order shear deformation plate theory SDPT, the basic assumptions for displacement behaviour[4-5] are

$$\bar{u}(x, y, z) = u(x, y) + z\psi_x(x, y)$$
$$\bar{v}(x, y, z) = v(x, y) + z\psi_y(x, y) \qquad (3.1)$$
$$\bar{w}(x, y, z) = w(x, y)$$

u, v and w are the mid-surface displacements in the x, y and z directions respectively. The displacement components with overbars are similar components at a general point. The quantities ψ_x and ψ_y are the rotations of the initial plate normals along the x and y directions respectively. The displacement at a general point is expressed in terms of five fundamental quantities, namely u, v, w, ψ_x and ψ_y.

Substitution of equations (3.1) in the Green's expression for in-plane nonlinear strains and using the usual von Karman assumptions for the strains,[25] gives the following expressions for five strain components at a general point:

$$\varepsilon = \begin{Bmatrix} \varepsilon_x \\ \varepsilon_y \\ \gamma_{yz} \\ \gamma_{zx} \\ \gamma_{xy} \end{Bmatrix} = \begin{Bmatrix} \dfrac{\partial u}{\partial x} + z\dfrac{\partial \psi_x}{\partial x} + \dfrac{1}{2}\left(\dfrac{\partial w}{\partial x}\right)^2 \\[2ex] \dfrac{\partial v}{\partial y} + z\dfrac{\partial \psi_y}{\partial y} + \dfrac{1}{2}\left(\dfrac{\partial w}{\partial y}\right)^2 \\[2ex] \dfrac{\partial w}{\partial y} + \psi_y \\[2ex] \dfrac{\partial w}{\partial x} + \psi_x \\[2ex] \dfrac{\partial u}{\partial y} + \dfrac{\partial v}{\partial x} + z\left(\dfrac{\partial \psi_x}{\partial y} + \dfrac{\partial \psi_y}{\partial x}\right) + \dfrac{\partial w}{\partial x}\dfrac{\partial w}{\partial y} \end{Bmatrix} \qquad (3.2)$$

The strain components ε_x, ε_y and γ_{xy} are the in-plane strains while γ_{yz} and γ_{zx} are the through-thickness engineering shear strains.

In general, laminated plate is composed of a number of bonded laminae with orthotropic material properties. On the assumption that each layer is in a state of plane stress, the stress- strain relationship at a general point in a layer is,

$$\begin{Bmatrix} \sigma_x \\ \sigma_y \\ \tau_{yz} \\ \tau_{zx} \\ \tau_{xy} \end{Bmatrix} = \begin{bmatrix} Q_{11} & & & & \\ Q_{12} & Q_{22} & & Symm & \\ 0 & 0 & Q_{44} & & \\ 0 & 0 & Q_{45} & Q_{55} & \\ Q_{16} & Q_{26} & 0 & 0 & Q_{66} \end{bmatrix} \begin{Bmatrix} \varepsilon_x \\ \varepsilon_y \\ \gamma_{yz} \\ \gamma_{zx} \\ \gamma_{xy} \end{Bmatrix}, \quad \sigma = Q\varepsilon \qquad (3.3)$$

where Q_{ij} $(i,j = 1,2,6)$ are the plane-stress stiffness coefficients and Q_{ij} $(ij = 4,5)$ are the through-thickness shear stiffness coefficients.

The constitutive equations for the laminate can be obtained through the use of equations (3.2) and (3.3) and integration through the thickness. The result is,

$$
\left\{
\begin{array}{c}
N_x \\
N_y \\
N_{xy} \\
M_x \\
M_y \\
M_{xy} \\
Q_y \\
Q_x
\end{array}
\right\}
= \int_{-\frac{h}{2}}^{\frac{h}{2}}
\left\{
\begin{array}{c}
\sigma_x \\
\sigma_y \\
\tau_{xy} \\
z\sigma_x \\
z\sigma_y \\
z\tau_{xy} \\
\tau_{yz} \\
\tau_{zx}
\end{array}
\right\} dx =
\tag{3.4}
$$

$$
\left[
\begin{array}{cccccccc}
A_{11} & & & & & & & \\
A_{12} & A_{22} & & & & & & \\
A_{16} & A_{26} & A_{66} & & \mathit{Symm} & & & \\
B_{11} & B_{12} & B_{16} & D_{11} & & & & \\
B_{12} & B_{22} & B_{26} & D_{12} & D_{22} & & & \\
B_{16} & B_{26} & B_{66} & D_{16} & D_{26} & D_{66} & & \\
0 & 0 & 0 & 0 & 0 & 0 & A_{44} & \\
0 & 0 & 0 & 0 & 0 & 0 & A_{45} & A_{45}
\end{array}
\right]
\left\{
\begin{array}{c}
\dfrac{\partial u}{\partial x} + \dfrac{1}{2}\left(\dfrac{\partial w}{\partial x}\right)^2 \\[2mm]
\dfrac{\partial v}{\partial y} + \dfrac{1}{2}\left(\dfrac{\partial w}{\partial y}\right)^2 \\[2mm]
\dfrac{\partial u}{\partial y} + \dfrac{\partial v}{\partial x} + \dfrac{\partial w}{\partial x}\dfrac{\partial w}{\partial y} \\[2mm]
\dfrac{\partial \psi_x}{\partial x} \\[2mm]
\dfrac{\partial \psi_y}{\partial y} \\[2mm]
\dfrac{\partial \psi_x}{\partial y} + \dfrac{\partial \psi_y}{\partial x} \\[2mm]
\dfrac{\partial w}{\partial y} + \psi_y \\[2mm]
\dfrac{\partial w}{\partial x} + \psi_x
\end{array}
\right\}
$$

N_x, N_y and N_{xy} are the membrane stress resultants per unit length. M_x, M_y and M_{xy} are the moments per unit length. Q_x and Q_y are the through-thickness shear forces per unit length. The laminate stiffness coefficients appearing in equations (3.4) are defined as

$$(A_{ij}, B_{ij}, D_{ij}) = \int_{-\frac{h}{2}}^{\frac{h}{2}} Q_{ij}(1, z, z^2) \ dz \ , \quad i, j = 1, 2, 6 \tag{3.5}$$

$$A_{ij} = k_i k_j \int_{-\frac{h}{2}}^{\frac{h}{2}} Q_{ij} \ dz \ , \quad i, j = 4, 5 \tag{3.6}$$

In the first-order shear deformation plate theory, the through-thickness shear strains are assumed to be uniform through the plate thickness. To allow for the fact that actual through-thickness shear strain distributions are not uniform through the plate thickness, shear correction factors k_4 and k_5 are introduced. Appropriate values of the shear correction factors can be determined using the method proposed by Whitney.[24]
The strain energy of a finite strip U_p is

$$U_p = \frac{1}{2} \int_{-\frac{b}{2}}^{\frac{b}{2}} \int_0^A \int_{-\frac{h}{2}}^{\frac{h}{2}} \sigma^T \varepsilon \ dz \ dx \ dy \tag{3.7}$$

Substitution of equations (3.2) and (3.3) into equation (3.7) and integration through the thickness, the strain energy of the finite strip can be put into the form of

$$U_p = \frac{1}{2} \iint \{a_1^T a_2^T\} \begin{Bmatrix} F_1 & F_3 \\ F_3 & F_2 \end{Bmatrix} \begin{Bmatrix} a_1 \\ a_2 \end{Bmatrix} dxdy + \frac{1}{2} \iint a_3^T H a_3 dxdy + $$
$$\frac{1}{6} \iint (b_1^T J_1 b_1 + b_2^T J_2 b_2) \ dxdy + \frac{1}{12} \iint c_1^T L_1 c_1 dxdy \tag{3.8}$$

Here

$$a_1 = \left\{ \frac{\partial u}{\partial x} \ \frac{\partial u}{\partial y} \ \frac{\partial v}{\partial x} \ \frac{\partial v}{\partial y} \right\}^T \tag{3.9}$$

$$a_2 = \left\{ \frac{\partial \psi_x}{\partial x} \ \frac{\partial \psi_x}{\partial y} \ \frac{\partial \psi_y}{\partial x} \ \frac{\partial \psi_y}{\partial y} \right\}^T \tag{3.10}$$

$$a_3 = \left\{ \psi_x \ \psi_y \ \frac{\partial w}{\partial x} \ \frac{\partial w}{\partial y} \right\}^T \tag{3.11}$$

$$b_1 = \left\{ \frac{\partial w}{\partial x} \quad \frac{\partial w}{\partial y} \quad \frac{\partial u}{\partial x} \quad \frac{\partial u}{\partial y} \quad \frac{\partial v}{\partial x} \quad \frac{\partial v}{\partial y} \right\}^T \tag{3.12}$$

$$b_2 = \left\{ \frac{\partial w}{\partial x} \quad \frac{\partial w}{\partial y} \quad \frac{\partial \psi_x}{\partial x} \quad \frac{\partial \psi_x}{\partial y} \quad \frac{\partial \psi_y}{\partial x} \quad \frac{\partial \psi_y}{\partial y} \right\}^T \tag{3.13}$$

$$c_1 = \left\{ \frac{\partial w}{\partial x} \quad \frac{\partial w}{\partial y} \right\}^T \tag{3.14}$$

$$F_1 = \left\{ \begin{matrix} A_{11} & & & Symm \\ A_{16} & A_{66} & & \\ A_{16} & A_{66} & A_{66} & \\ A_{12} & A_{26} & A_{26} & A_{22} \end{matrix} \right\} \tag{3.15}$$

$$F_2 = \left\{ \begin{matrix} D_{11} & & & Symm \\ D_{16} & D_{66} & & \\ D_{16} & D_{66} & D_{66} & \\ D_{12} & D_{26} & D_{26} & D_{22} \end{matrix} \right\} \tag{3.16}$$

$$F_3 = \left\{ \begin{matrix} B_{11} & & & Symm \\ B_{16} & B_{66} & & \\ B_{16} & B_{66} & B_{66} & \\ B_{12} & B_{26} & B_{26} & B_{22} \end{matrix} \right\} \tag{3.17}$$

$$H = \left\{ \begin{matrix} A_{55} & & & Symm \\ A_{45} & A_{44} & & \\ A_{55} & A_{45} & A_{55} & \\ A_{45} & A_{44} & A_{45} & A_{44} \end{matrix} \right\} \tag{3.18}$$

$$J_1 = \left\{ \begin{array}{ccc} \left\{ \begin{array}{l} A_{11}\dfrac{\partial u}{\partial x} + A_{12}\dfrac{\partial v}{\partial y} + \\[2mm] A_{16}\left(\dfrac{\partial u}{\partial y} + \dfrac{\partial v}{\partial x}\right) + B_{11}\dfrac{\partial \psi_x}{\partial x} + \\[2mm] B_{12}\dfrac{\partial \psi_y}{\partial y} + B_{16}\left(\dfrac{\partial \psi_x}{\partial y} + \dfrac{\partial \psi_y}{\partial x}\right) \end{array} \right\} & & \text{Symm} \\[14mm]

\left\{ \begin{array}{l} A_{16}\dfrac{\partial u}{\partial x} + A_{26}\dfrac{\partial v}{\partial y} + \\[2mm] A_{66}\left(\dfrac{\partial u}{\partial y} + \dfrac{\partial v}{\partial x}\right) + B_{16}\dfrac{\partial \psi_x}{\partial x} + \\[2mm] B_{26}\dfrac{\partial \psi_y}{\partial y} + B_{66}\left(\dfrac{\partial \psi_x}{\partial y} + \dfrac{\partial \psi_y}{\partial x}\right) \end{array} \right\} & \left\{ \begin{array}{l} A_{12}\dfrac{\partial u}{\partial x} + A_{22}\dfrac{\partial v}{\partial y} + \\[2mm] A_{26}\left(\dfrac{\partial u}{\partial y} + \dfrac{\partial v}{\partial x}\right) + B_{12}\dfrac{\partial \psi_x}{\partial x} + \\[2mm] B_{22}\dfrac{\partial \psi_y}{\partial y} + B_{26}\left(\dfrac{\partial \psi_x}{\partial y} + \dfrac{\partial \psi_y}{\partial x}\right) \end{array} \right\} & \\[14mm]

A_{11}\dfrac{\partial w}{\partial x} + A_{16}\dfrac{\partial w}{\partial y} & A_{16}\dfrac{\partial w}{\partial x} + A_{12}\dfrac{\partial w}{\partial y} & 0 \\[4mm]
A_{16}\dfrac{\partial w}{\partial x} + A_{66}\dfrac{\partial w}{\partial y} & A_{66}\dfrac{\partial w}{\partial x} + A_{26}\dfrac{\partial w}{\partial y} & 0 \quad 0 \\[4mm]
A_{16}\dfrac{\partial w}{\partial x} + A_{66}\dfrac{\partial w}{\partial y} & A_{66}\dfrac{\partial w}{\partial x} + A_{26}\dfrac{\partial w}{\partial y} & 0 \quad 0 \quad 0 \\[4mm]
A_{12}\dfrac{\partial w}{\partial x} + A_{26}\dfrac{\partial w}{\partial y} & A_{26}\dfrac{\partial w}{\partial x} + A_{22}\dfrac{\partial w}{\partial y} & 0 \quad 0 \quad 0 \quad 0
\end{array} \right\}$$

(3.19)

$$J_2 = \left\{ \begin{array}{ccccc}
0 & & & & \\[3mm]
0 & 0 & & \text{Symm} & \\[3mm]
B_{11}\dfrac{\partial w}{\partial x} + B_{16}\dfrac{\partial w}{\partial y} & B_{16}\dfrac{\partial w}{\partial x} + B_{12}\dfrac{\partial w}{\partial y} & 0 & & \\[4mm]
B_{16}\dfrac{\partial w}{\partial x} + B_{66}\dfrac{\partial w}{\partial y} & B_{66}\dfrac{\partial w}{\partial x} + B_{26}\dfrac{\partial w}{\partial y} & 0 & 0 & \\[4mm]
B_{16}\dfrac{\partial w}{\partial x} + B_{66}\dfrac{\partial w}{\partial y} & B_{66}\dfrac{\partial w}{\partial x} + B_{26}\dfrac{\partial w}{\partial y} & 0 & 0 & 0 \\[4mm]
B_{12}\dfrac{\partial w}{\partial x} + B_{26}\dfrac{\partial w}{\partial y} & B_{26}\dfrac{\partial w}{\partial x} + B_{22}\dfrac{\partial w}{\partial y} & 0 & 0 & 0 \quad 0
\end{array} \right\}$$

(3.20)

$$
L_1 = \left\{
\begin{array}{ll}
\left\{
\begin{array}{l}
\dfrac{3}{2} A_{11}\left(\dfrac{\partial w}{\partial x}\right)^2 + 3A_{16}\dfrac{\partial w}{\partial x}\dfrac{\partial w}{\partial y} + \\[2mm]
\left(A_{66} + \dfrac{1}{2}A_{12}\right)\left(\dfrac{\partial w}{\partial y}\right)^2
\end{array}
\right\}
& \qquad Symm \\[8mm]
\left\{
\begin{array}{l}
\dfrac{3}{2} A_{16}\left(\dfrac{\partial w}{\partial x}\right)^2 + \dfrac{3}{2} A_{26}\left(\dfrac{\partial w}{\partial y}\right)^2 + \\[2mm]
(A_{12} + 2A_{66})\dfrac{\partial w}{\partial x}\dfrac{\partial w}{\partial y}
\end{array}
\right\}
&
\left\{
\begin{array}{l}
3A_{26}\dfrac{\partial w}{\partial x}\dfrac{\partial w}{\partial y} + \dfrac{3}{2} A_{22}\left(\dfrac{\partial w}{\partial y}\right)^2 + \\[2mm]
\left(A_{66} + \dfrac{1}{2}A_{12}\right)\left(\dfrac{\partial w}{\partial y}\right)^2
\end{array}
\right\}
\end{array}
\right.
$$

$$(3.21)$$

In the expression for U_p in equation (3.8), the strain energy can be separated into three categories according to the order of nonlinearity. Integral with a factor of 1/2 in front represents quadratic energy related to in-plane stretching-shearing action, out-of-plane bending-twisting action and coupling between these two actions arising from material effects. Integral with a factor of 1/6 in front represents cubic energy related to the coupling between in-plane stretching-shearing and out-of-plane bending-twisting actions, arising both from the interaction between the nonlinear terms in w and the linear terms in the strain-displacement equations and from material effects. Integral with a factor of 1/12 in front represents quartic energy arising from the nonlinear terms in w alone.

For plates subjected to prescribed end shortening strain, the total potential energy π_p of the associated end load will thus be independent of any degree of freedom of the system. Thus,

$$\pi_p = U_p \qquad (3.22)$$

Solution of the nonlinear problem is sought through the application of the principle of minimum potential energy once displacement field representing the variations of the fundamental quantities are identified. The displacement field selected for the typical SDPT finite strip are

$$u = \varepsilon\left(\frac{A}{2} - x\right) + \sum_{i=1}^{ru} U_i(x)g_i^u(y) \qquad (3.23)$$

$$v = \alpha\varepsilon\, y + \sum_{i=0}^{rv} V_i(x)g_i^v(y) \qquad (3.24)$$

$$w = \sum_{i=1}^{rw} W_i(x) g_i^{w}(y) \tag{3.25}$$

$$\psi_y = \sum_{i=1}^{rw} \psi_{yi}(x) g_i^{\psi_y}(y) \tag{3.26}$$

$$\psi_x = \sum_{i=1}^{rw} \psi_{xi}(x) g_i^{\psi_x}(y) \tag{3.27}$$

ε denotes the prescribed end shortening strain and α is a constant. ru, rv and rw are the highest terms used in the corresponding longitudinal series terms. Same number of terms is used for w, ψ_y and ψ_x so as to be in consistent with the definition of the strains in equations (3.2). The longitudinal functions $U_i(x)$, $V_i(x)$, $W_i(x)$, $\psi_{yi}(x)$ and $\psi_{xi}(x)$ are taken to be trigonometric functions satisfying the kinematic conditions at the strip ends, which are assumed to be restrained by rigid platens such that uniform end shortening occurs. The various crosswise functions $g_i(y)$ are polynomial interpolation functions with undetermined displacement unknowns (degrees of freedom).

It is assumed that the plate is simply supported for out-of-plane behaviour at its loaded ends. The longitudinal series terms in the expressions for w, $\psi_y(x)$ and $\psi_x(x)$ are

$$W_i(x) = \psi_{yi}(x) = \sin\left(\frac{i\pi x}{A}\right), \quad \psi_{xi}(x) = \cos\left(\frac{i\pi x}{A}\right) \tag{3.28}$$

The uniform end shortening is exemplified by the presence of the prescribed end shortening strain ε in the x-direction and the longitudinal series terms for u are

$$U_i(x) = \sin\left(\frac{i\pi x}{A}\right) \tag{3.29}$$

The longitudinal series terms chosen for v depend upon the kinematic conditions at the loaded ends and two types of problem can be identified. The loaded ends can be either free to expand in-plane (problem Type A) or is restrained completely against such expansion (problem Type B). These are the two extreme situations of lateral boundary conditions representing frictionless platens and fully frictional platens respectively.

Problem Type A. The longitudinal series terms for v are

$$V_i(x) = \cos\left(\frac{i\pi x}{A}\right) \tag{3.30}$$

The term $\alpha \varepsilon y$ represents the lateral expansion of a flat unbuckled plate under uniform end compression so that a trivial primary equilibrium path is invoked without involving any unprescribed degree of freedom. This situation applies for isotropic plates (where $\alpha = v$) and for orthotropic, symmetrically laminated plates (where $\alpha = A_{12}/A_{22}$). For laminated plates involving anisotropy and coupling of in-plane and out-of-plane properties through material effects, the primary path is non-trivial. In this case, the presence of the term $\alpha \varepsilon y$ is irrational. For consistency, this term is retained with $\alpha = A_{12}/A_{22}$ in the analysis of the full range of plate properties, though its presence is not necessary.

Problem Type B. The kinematic condition $v = 0$ applies at the loaded ends. The longitudinal series terms for v are

$$V_i(x) = \sin\left(\frac{i\pi x}{A}\right) \tag{3.31}$$

with the $i = 0$ term disappears and $\alpha = 0$.

The crosswise functions $g(y)$ for both types of problem defined above are identical. For each longitudinal series term i,

$$g^u(y) = \sum_{m=1}^{M} N_m(y) u_m \tag{3.32}$$

$$g^v(y) = \sum_{m=1}^{M} N_m(y) v_m \tag{3.33}$$

$$g^w(y) = \sum_{m=1}^{M} N_m(y) w_m \tag{3.34}$$

$$g^{\psi_y}(y) = \sum_{m=1}^{M} N_m(y) \psi_{ym} \tag{3.35}$$

$$g^{\psi_x}(y) = \sum_{m=1}^{M} N_m(y) \psi_{xm} \tag{3.36}$$

$N_m(x)$ are crosswise interpolation functions and U_m, etc., are the reference line values of u, etc.. In the present study, same crosswise interpolation functions are used for each of the five fundamental quantities. Allowance is made for three kinds of strip model corresponding to linear ($M = 2$),

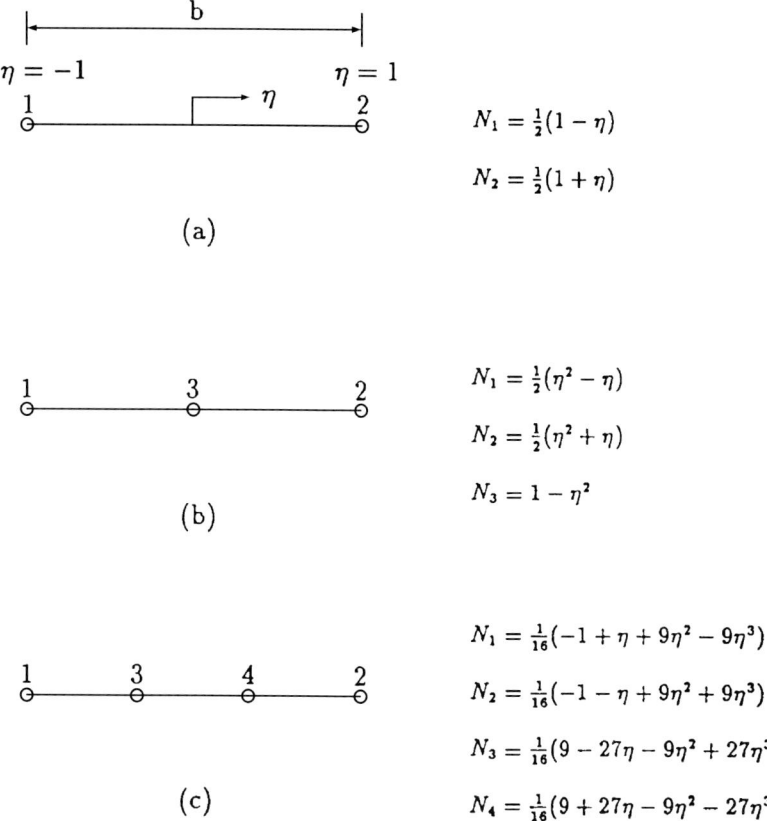

$$N_1 = \tfrac{1}{2}(1 - \eta)$$

$$N_2 = \tfrac{1}{2}(1 + \eta)$$

(a)

$$N_1 = \tfrac{1}{2}(\eta^2 - \eta)$$

$$N_2 = \tfrac{1}{2}(\eta^2 + \eta)$$

$$N_3 = 1 - \eta^2$$

(b)

$$N_1 = \tfrac{1}{16}(-1 + \eta + 9\eta^2 - 9\eta^3)$$

$$N_2 = \tfrac{1}{16}(-1 - \eta + 9\eta^2 + 9\eta^3)$$

$$N_3 = \tfrac{1}{16}(9 - 27\eta - 9\eta^2 + 27\eta^3)$$

$$N_4 = \tfrac{1}{16}(9 + 27\eta - 9\eta^2 - 27\eta^3)$$

(c)

Figure 2. SDPT finite strip models corresponding to (a) linear, (b) quadratic and (c) cubic crosswise interpolation.

quadratic ($M = 3$) and cubic ($M = 4$) crosswise interpolation functions. The degrees of freedom at each reference line are values of u, v, w, ψ_x and ψ_y. Figure 2 shows the end views of the three kinds of strip model. The reference lines are numbered with the corresponding shape functions defined in terms of the non-dimensional co-ordinate $\eta = 2y/b$.

The potential energy of a finite strip can now be determined by incorporating the definitions of the displacement functions. All integrations in the x-direction related to the trigonometric terms are evaluated analytically. Integrations in the y-direction are determined numerically using Gauss Quadrature. The concepts of full, selective and reduced integration widely used in the finite element analysis of shear-deformable plates have been discussed by Dawe and Azizian[26] within the context of nonlinear SDPT finite strips. Results of numerical studies concerning the large deflection of isotropic plates indicate that the use of selective or reduced integration can lead to improved efficiency of solution for strip models based on linear or quadratic crosswise interpolation functions. Despite this, for the purpose of this study, all the numerical results for the SDPT finite strip method are based on the use of sufficient Gauss points to ensure the exact calculation of all stiffness terms.

5.4. NONLINEAR FORMULATION BASED ON CPT

The classical plate theory is based on the Kirchhoff normalcy condition in neglecting the through-thickness shear strains in equations (3.2), i.e.

$$
\gamma_{yz} = \frac{\partial w}{\partial y} + \psi_y = 0 \Rightarrow \psi_y = -\frac{\partial w}{\partial y}
$$
$$
\gamma_{zx} = \frac{\partial w}{\partial y} + \psi_x = 0 \Rightarrow \psi_x = -\frac{\partial w}{\partial y}
$$
(4.1)

Equations (4.1) are substituted into equations (3.1) so that the displacement at a general point is now expressed in terms of only three fundamental mid-surface quantities, namely u, v and w, rather than the five of SDPT. Substitution of equation (4.1) into equations (3.2) to (3.4) and (3.8) in turn to produce a strain energy for CPT in the form of

$$
U_p = \frac{1}{2} \iint \{a_1{}^T a_4{}^T\} \begin{Bmatrix} F_1 & F_3 \\ F_3 & F_2 \end{Bmatrix} \begin{Bmatrix} a_1 \\ a_4 \end{Bmatrix} dxdy +
$$
$$
\frac{1}{6} \iint (b_1{}^T J_3 b_1 + b_3{}^T J_2 b_3) \, dxdy + \frac{1}{12} \iint c_1{}^T L_1 c_1 dxdy
$$
(4.2)

The matrices a_1, F_1, F_2, b_1, J_2, c_1 and L_1 are the same as those defined in Section 3. The new matrices a_4, b_3 and J_3 are

$$
a_4 = \left\{ -\frac{\partial^2 w}{\partial x^2} \quad -\frac{\partial^2 w}{\partial x \partial y} \quad \frac{\partial^2 w}{\partial x \partial y} \quad -\frac{\partial^2 w}{\partial y^2} \right\}^T
$$
(4.3)

$$
b_3 = \left\{ \frac{\partial w}{\partial x} \quad \frac{\partial w}{\partial y} \quad -\frac{\partial^2 w}{\partial x^2} \quad \frac{\partial^2 w}{\partial x \partial y} \quad \frac{\partial^2 w}{\partial x \partial y} \quad -\frac{\partial^2 w}{\partial y^2} \right\}^T
$$
(4.4)

$$J_3 = \left\{ \begin{array}{l}
\left\{ A_{11}\dfrac{\partial u}{\partial x} + A_{12}\dfrac{\partial v}{\partial y} + A_{16}\left(\dfrac{\partial u}{\partial y}+\dfrac{\partial v}{\partial x}\right) - B_{11}\dfrac{\partial^2 w}{\partial x^2} - B_{12}\dfrac{\partial^2 w}{\partial y^2} - 2B_{16}\dfrac{\partial^2 w}{\partial x\partial y} \right\} \qquad\qquad Symm \\[2em]
\left\{ A_{16}\dfrac{\partial u}{\partial x} + A_{26}\dfrac{\partial v}{\partial y} + A_{66}\left(\dfrac{\partial u}{\partial y}+\dfrac{\partial v}{\partial x}\right) - B_{16}\dfrac{\partial^2 w}{\partial x^2} - B_{26}\dfrac{\partial^2 w}{\partial y^2} - 2B_{66}\dfrac{\partial^2 w}{\partial x\partial y} \right\} \quad \left\{ A_{12}\dfrac{\partial u}{\partial x} + A_{22}\dfrac{\partial v}{\partial y} + A_{26}\left(\dfrac{\partial u}{\partial y}+\dfrac{\partial v}{\partial x}\right) - B_{12}\dfrac{\partial^2 w}{\partial x^2} - B_{22}\dfrac{\partial^2 w}{\partial y^2} - 2B_{26}\dfrac{\partial^2 w}{\partial x\partial y} \right\} \\[2em]
\begin{array}{llcccc}
A_{11}\dfrac{\partial w}{\partial x} + A_{16}\dfrac{\partial w}{\partial y} & A_{16}\dfrac{\partial w}{\partial x} + A_{12}\dfrac{\partial w}{\partial y} & 0 & & & \\[1em]
A_{16}\dfrac{\partial w}{\partial x} + A_{66}\dfrac{\partial w}{\partial y} & A_{66}\dfrac{\partial w}{\partial x} + A_{26}\dfrac{\partial w}{\partial y} & 0 & 0 & & \\[1em]
A_{16}\dfrac{\partial w}{\partial x} + A_{66}\dfrac{\partial w}{\partial y} & A_{66}\dfrac{\partial w}{\partial x} + A_{26}\dfrac{\partial w}{\partial y} & 0 & 0 & 0 & \\[1em]
A_{12}\dfrac{\partial w}{\partial x} + A_{26}\dfrac{\partial w}{\partial y} & A_{26}\dfrac{\partial w}{\partial x} + A_{22}\dfrac{\partial w}{\partial y} & 0 & 0 & 0 & 0
\end{array}
\end{array} \right.$$

$$(4.5)$$

This strain energy expression closely resembles the corresponding SDPT based expression for U_p in Section 3. The second integral on the right hand side of equation (3.8) representing the through-thickness shear strain energy is now absent in the CPT formulation.

As in the SDPT based formulation, the total potential energy π_p is identical to the strain energy U_p. The number of fundamental quantities over the plate mid surface is reduced from five (in the SDPT formulation) to three, the u, v and w respectively. The displacement field selected for the CPT finite strip are the same as those defined in equations (3.23) to (3.25) and the longitudinal series defined in equations (3.28) to (3.31) are also applicable to the CPT finite strip.

Identical crosswise functions are used to represent the displacement fields u and v and allowance is made for linear ($M = 2$), quadratic ($M = 3$) and cubic ($M = 4$) variation across a strip. For each longitudinal series term i,

$$g^u(y) = \sum_{m=1}^{M} N_m^u(y)\, u_m \qquad (4.6)$$

$$g^v(y) = \sum_{m=1}^{M} N_m^v(y)\, v_m \qquad\qquad (4.7)$$

$$g^w(y) = N_1^w(y)\, w_1 + N_2^w(y)\, \psi_1 + N_3^w(y)\, w_2 + N_4^w(y)\, \psi_2 \qquad (4.8)$$

where $\psi = \partial w/\partial y$. For $M = 2$ the references lines are located only at the outside edges of the finite strip, whilst for $M = 3$ and $M = 4$ there are one or two additional reference lines inside the strip. Figure 3 shows the end views of the three kinds of strip model. The degrees of freedom u and v appear in all the reference lines whereas w and ψ only exist at the two outside reference lines. The shape functions occur in Figure 3 are defined in terms of the non-dimensional co-ordinate $\eta = 2y/b$.

In evaluating π_p, all integrations in the x-direction are determined analytically, as for the SDPT approach. Integrations in the y-direction are determined numerically using Gauss Quadrature and sufficient Gauss points have been used so that the integrations are effectively exact.

5.5. BOUNDARY CONDITIONS

The kinematic boundary conditions are satisfied explicitly at the loaded ends of a plate using the defined displacement field. In general, the natural boundary conditions are not satisfied in the case of anisotropic plates, i.e. laminated plates in non-symmetric arrangement. In particular, one natural boundary condition that should apply at the plate ends is that the bending moment M_x be zero, whist at the same time the twisting moment M_{xy} be non-zero. For anisotropic material, it is not possible to meet these requirements simultaneously when using the trigonometric series displacement field. Instead the desired condition $Mx = 0$ is replaced by

$$\frac{\partial \psi_x}{\partial x} = 0 \quad \text{at} \quad x = 0, A \qquad\qquad (5.1)$$

and this introduces some degree of constraint to the problem. However, for moderate anisotropy the effect would be expected to be quite small as demonstrated in[19] when comparison is made with finite element results. At the longitudinal edges any desired kinematic boundary condition can be prescribed in the usual manner, while any natural boundary condition can be satisfied as a result of the variational procedure.

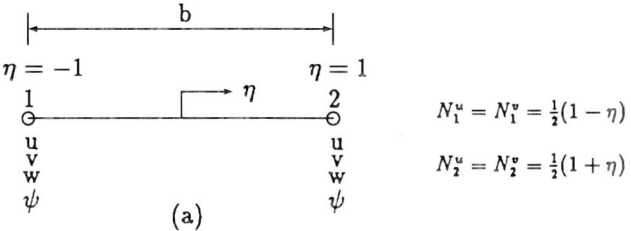

$$N_1^u = N_1^v = \tfrac{1}{2}(1 - \eta)$$

$$N_2^u = N_2^v = \tfrac{1}{2}(1 + \eta)$$

(a)

$$N_1^u = N_1^v = \tfrac{1}{2}(\eta^2 - \eta)$$

$$N_2^u = N_2^v = \tfrac{1}{2}(\eta^2 + \eta)$$

$$N_3^u = N_3^v = 1 - \eta^2$$

(b)

$$N_1^u = N_1^v = \tfrac{1}{16}(-1 + \eta + 9\eta^2 - 9\eta^3)$$

$$N_2^u = N_2^v = \tfrac{1}{16}(-1 - \eta + 9\eta^2 + 9\eta^3)$$

$$N_3^u = N_3^v = \tfrac{1}{16}(9 - 27\eta - 9\eta^2 + 27\eta^3)$$

$$N_4^u = N_4^v = \tfrac{1}{16}(9 + 27\eta - 9\eta^2 - 27\eta^3)$$

(c)

For (a), (b) and (c) :

$$N_1^w = \tfrac{1}{4}(2 - 3\eta + \eta^3) \qquad N_2^w = \tfrac{b}{8}(1 - \eta - \eta^2 + \eta^3)$$

$$N_3^w = \tfrac{1}{4}(2 + 3\eta - \eta^3) \qquad N_4^w = \tfrac{b}{8}(-1 - \eta + \eta^2 + \eta^3)$$

Figure 3. CPT finite strip models corresponding to representation of u and v by (a) linear, (b) quadratic and (c) cubic crosswise interpolation.

5.6. SOLUTION PROCEDURE

Substitution of the chosen displacement field into the expressions for strain energy, the potential energy of a finite strip can be determined as a function of the strip degrees of freedom at the reference lines for all the longitudinal series terms. For a whole plate, comprising an assembly of finite strips, the total potential energy is simply the summation of the potential energies of the individual finite strips. The potential energy of a whole plate can be expressed as

$$\pi_p = -\varepsilon d^T v + \frac{1}{2} d^T (K - \varepsilon K^*) d + \frac{1}{6} d^T K_1 d + \frac{1}{12} d^T K_2 d \qquad (6.1)$$

The matrices K, K_1, K_2 and K^* are square and symmetric. The coefficients of K and K^* are constants whilst those of K_1 and K_2 are linear and quadratic functions, respectively, of the displacements. d is a column vector of strip degrees of freedom. V is a column vector of constants (which becomes a null vector in problem Type A for isotropic plates or symmetric orthotropic laminates).

A set of nonlinear plate equilibrium equations corresponding to a specific value of ε is then obtained by applying the principle of minimum potential energy with respect to the column vector d

$$-\varepsilon V + \left(K - \varepsilon K^* + \frac{1}{2} K_1 + \frac{1}{3} K_2 \right) d = 0 \qquad (6.2)$$

These set of nonlinear equations can be solved using an iterative procedure. In this study the Newton-Raphson method is chosen. Assume d_i is an approximate trial solution at a particular prescribed end strain ε at iteration i. The tangent stiffness matrix K_T defined as

$$K_T = K - \varepsilon K^* + K_1 + K_2 \qquad (6.3)$$

can be evaluated based on the approximate trial solution d_i. The correction δd_i can be obtained by

$$\delta d_i = K_T^{-1} \left\{ \varepsilon V - \left(K - \varepsilon K^* + \frac{1}{2} K_1 + \frac{1}{3} K_2 \right) d_i \right\} \qquad (6.4)$$

The displacement vector d is updated at the end of every iteration using

$$d_{i+1} = d_i + \delta d_i \qquad (6.5)$$

The iterative process is repeated until the convergence criterion

$$\sqrt{\frac{\delta d_i^T \delta d_i}{d_i^T d_i}} \leq 0.0005 \qquad (6.6)$$

is satisfied.

Table 1. End and edge conditions for five isotropic square plates.

Case	In-plane displacement conditions		Out-of-plane displacement conditions	
	Loaded ends	Unloaded edges	Loaded ends	Unloaded edges
1	Free	Straight	SS	SS
2	Free	Free	SS	SS
3	Free	Straight	SS	C
4	Held	Free	SS	SS
5	Held	Held	SS	SS

Notes: For the in-plane conditions at the loaded ends, Free and Held denote whether the ends are free to expand laterally or held against expansion. At the unloaded edges, Free denotes that the edges are free to move laterally, Straight that the edges can move laterally but are kept straight and Held that the edges are held against any lateral movement whatsoever. For the out-of-plane conditions, SS and C denote simply supported and clamped ends/edges respectively.

One particular quantity of interest is the average longitudinal force N_{av} acting on the plate. This is determined by

$$N_{av} = \frac{1}{A} \sum \int_0^b \int_0^A N_x(x,y) \, dx \, dy \qquad (6.7)$$

where the summation relates to all the individual strips.

5.7. APPLICATIONS

5.7.1. Isotropic Square Plates — CPT

Table 1 describes five isotropic square plates with different boundary conditions at the loaded ends, where uniform end shortening is imposed, and at the unloaded edges. Cases 1, 2 and 3 are of the problem Type A category and were considered by Yamaki[27] while cases 4 and 5 are of the problem Type B category, not considered by Yamaki. The Poisson's ratio is 1/3 and A/h is 120.

In the present approach half of a plate is modelled by four CPT finite strips having quadratic membrane displacement approximation ($M = 3$). The series terms used for u are $i = 2,4,6$ and for w are $i = 1,3$. For problem Type A v is represented by the cosine series $i = 0,2,4,6$ while for problem Type B the sine series $i = 1,3,5,7$ is used instead.

The finite strip results for the five different cases are shown graphically in Figure 4, giving the load factor — relative end strain history and the load factor — maximum deflection history. The load factor F is defined as

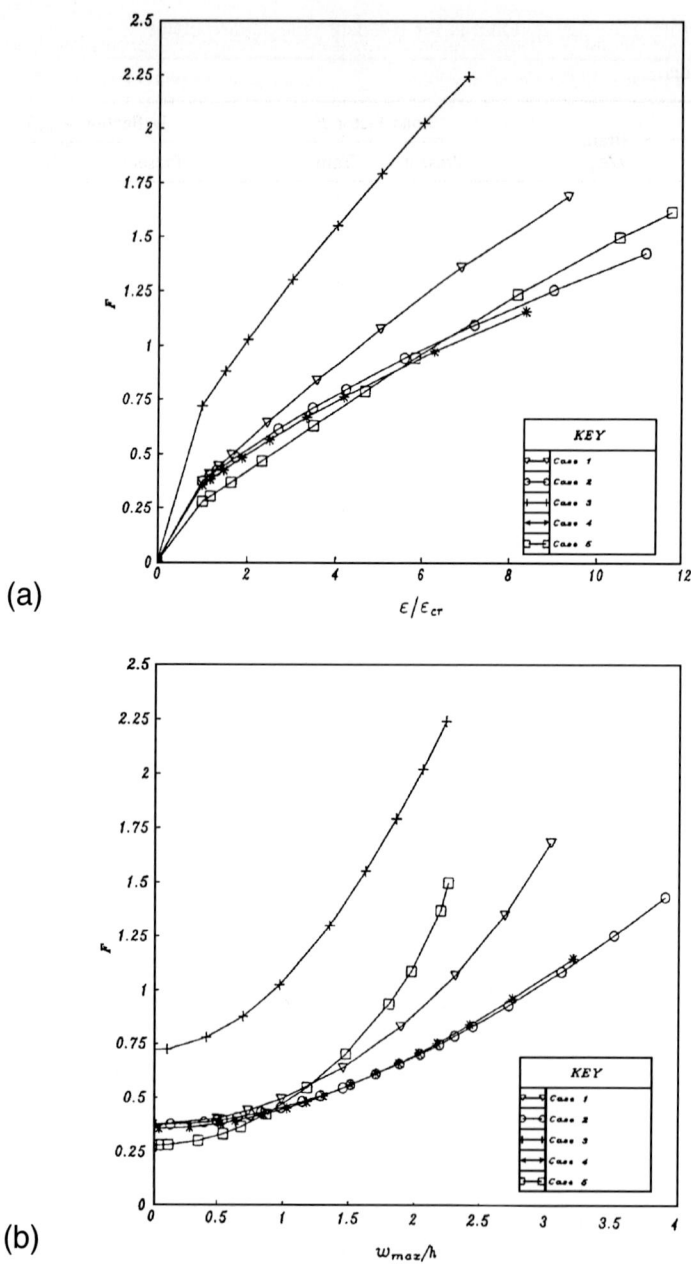

Figure 4. (a) Longitudinal force — end shortening behaviour for five isotropic square plates. (b) Longitudinal force — central deflection behaviour for five isotropic square plates.

Table 2. Comparison of finite strip results with those of Yamaki[27] for two isotropic square plates.

Case	Strain $\varepsilon/\varepsilon_{cr}$	Load factor F		Deflection w_{max}/h	
		Present	Yamaki	Present	Yamaki
1	1.0000	0.3750	0.375		
	1.1160	0.4060	0.406	0.4977	0.4979
	1.6590	0.4971	0.497	0.9838	0.9840
	2.4684	0.6431	0.643	1.450	1.450
	3.5820	0.8394	0.839	1.891	1.892
	5.0464	1.079	1.079	2.301	2.302
	6.8987	1.363	1.364	2.680	2.682
	9.3910	1.696	1.703	3.026	3.037
2	1.0000	0.3750	0.375		
	1.1392	0.3961	0.396	0.4960	0.4975
	1.5520	0.4568	0.457	0.9819	0.9813
	2.2166	0.5488	0.549	1.445	1.444
	3.1169	0.6643	0.664	1.884	1.884
	4.2510	0.7977	0.798	2.307	2.304
	5.6151	0.9437	0.945	2.714	2.711
	7.2123	1.099	1.103	3.112	3.110
	9.0526	1.263	1.273	3.506	3.506
	11.138	1.434	1.456	3.900	3.904

$$F = \frac{N_{av} A}{\pi^2 E h^3} \qquad (7.1)$$

The relative end strain is $\varepsilon/\varepsilon_{cr}$, where ε_{cr} is the end shortening strain at buckling.

For Cases 1 and 2 typical values of load factors and maximum deflections at relative end strains computed by the CPT finite strip are compared with those of Yamaki. These are included in Table 2 and in general the two sets of results compare very closely.

5.7.2. Isotropic Square Plate - SDPT and CPT

The isotropic square plate is of side length A and $A/h = 20$. Poisson's ratio has the value 1/3 and $k_4^2 = k_5^2 = 5/6$. The unloaded edges of the plate are clamped against out-of-plane movement (whilst the loaded ends are simply supported, of course). In the plane of the plate, lateral expansion is prevented at the loaded ends (problem Type B) and the unloaded edges are free to wave.

The problem is analysed in the context of SDPT using both the finite strip and finite element method and in CPT using the finite strip method. The

Table 3. Comparison of SDPT finite strip (FSM) and finite element (FEM) results for the isotropic square plate.

		Critical strain		Post-buckling stiffness	
			Relative computation		Relative computation
Type of analysis	n	ε_{cr}	time (%)	S^*/S	time (%)
FSM, $M = 3$	1	0.02953	0.4	0.404	0.3
	2	0.01905	0.9	0.609	0.5
	4	0.01738	1.7	0.574	1.2
	6	0.01723	2.6	0.570	1.6
	8	0.01720	3.7	0.570	2.1
FSM, $M = 4$	1	0.01748	0.8	0.606	0.4
	2	0.01720	1.5	0.571	0.8
	4	0.01719	3.0	0.569	1.6
	6	0.01719	4.3	0.569	2.3
FEM	1	0.01852	0.5	0.592	0.9
	2	0.01728	2.8	0.579	4.9
	4	0.01720	14.6	0.588	19.5
	6	0.01721	40.3	0.568	49.4
	8	0.01721	100.0	0.568	100.0

finite element used in this study is based on the heterosis element of Hughes and Cohen[28] extended to include a membrane displacement field and to embrace geometric nonlinearity.[19] In the finite strip approach half of the plate is modelled by n strips of either quadratic ($M = 3$) or cubic ($M = 4$) kind. The series terms used are sin2,4,6, sin1,3,5,7, sin1,3, cos1,3 and sin1,3 for u, v, w, ψ_x and ψ_y respectively. In the finite element approach, a double-symmetric quarter of the plate is modelled with a regular mesh of $n \times n$ equal elements.

Table 3 compares the values of the buckling strains ε_{cr} and the immediate post-buckling relative in-plane stiffness S^*/S using SDPT finite strip and finite element for various values of n. Results from the two methods converge to closely comparable common values of ε_{cr} and S^*/S with increasing n. Also recorded in Table 3 is the relative time of computation for the two methods as a percentage of the time taken using the finest finite element mesh based on the heterosis element. It is clear that the finite strip method is more economical than the finite element method. The performance of the two methods at particular post-buckling levels has also been studied in.[19] The economy of the finite strip method relative to the finite element method is again apparent. A significant part of the reason for this is that in the finite

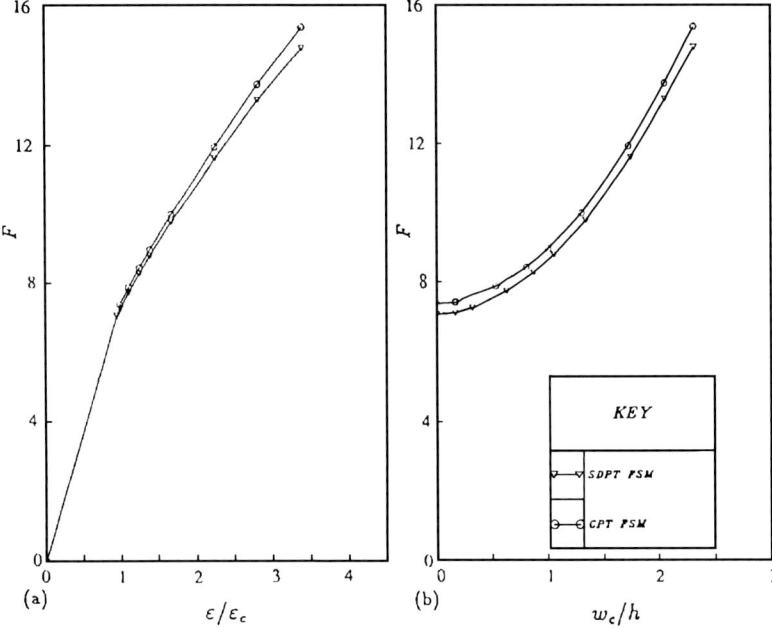

Figure 5. (a) Longitudinal force — end shortening behaviour of moderate thick isotropic square plate. (b) Longitudinal force — central deflection behaviour of moderate thick isotropic square plate.

element method the bandwidth of the system equations is dependent on the mesh size, whereas it is not so in the finite strip method.

Figure 5 shows the post-buckling behaviour of the moderately thick isotropic plate for the variations of average end force with end shortening strain and with central deflection, based on the SDPT and CPT finite strip (both using eight quadratic strips in half of the plate).

5.7.3. Square Unsymmetrical Cross-Ply Laminates

The laminates are constructed of material having the following properties

$$\frac{E_L}{E_T} = 40 \ , \ \frac{G_{LT}}{E_T} = 0.5 \ , \ \frac{G_{TT}}{E_T} = 0.6 \ , \ \nu_{LT} = 0.25$$

and are made of an even number of equal-thickness layers at 0/90/0/90/0/90/... arrangement. The laminates are simply supported all round for out-of-plane deflection. In-plane lateral expansion is allowed at the loaded ends and

the unloaded edges are free to expand laterally, i.e. free to wave. The
analysis is of the Type A category.

For cross-ply laminates, the non-zero coupling coefficients are B_{11} and
B_{22}, with $B_{22} = -B_{11}$. The presence of the B_{11} and B_{22} coefficients implies
that out-of-plane behaviour will commence from the beginning of the application of the in-plane loading and hence no buckling load exists as such,
except for the limiting situation of an infinite number of layers.

Performance of the three kinds of CPT finite strips, namely linear, quadratic and cubic finite strips, was studied in.[18] The results have shown clearly
that there is marked advantage in using the quadratic and cubic strips rather

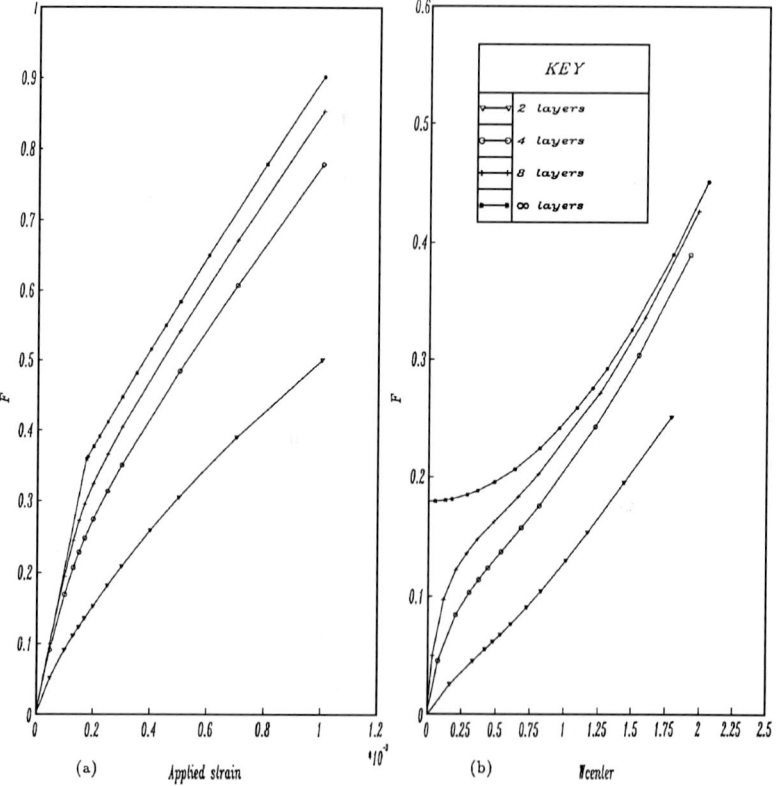

Figure 6.   (a) Longitudinal force — end shortening behaviour with different number
of layers for cross-ply square laminates. (b) Longitudinal force — central deflection
behaviour with different number of layers for cross-ply square laminates.

Table 4. Comparison of SDPT and CPT results for the eight-layer cross-ply square laminates.

A/h	Percentage strain	Type of analysis	W_c/h	F
20	1.00	SDPT	1.074	45.16
		CPT	1.050	47.35
	2.50	SDPT	1.945	78.16
		CPT	1.972	85.02
100	0.05	SDPT	1.253	53.98
		CPT	1.252	54.08
	0.10	SDPT	1.972	84.82
		CPT	1.972	85.02

than the basic linear strips, especially at relatively large end shortening strain. Therefore, quadratic strip will be used in this study. Following the recommendation in[18], the longitudinal series terms used for u, v, w, ψ_x and ψ_y are sin2,4,6, cos0,2,4,6, sin1,3, cos1,3 and sin1,3 respectively.

The complete in-plane force — end shortening strain and in-plane — central deflection variations for strains up to 0.1% with $A/h = 100$ are shown in Figure 6 for the cross-ply laminates with 2, 4, 8 and an infinite number of layers (with $B_{ij} = 0$ representing an infinite number of layers). These graphs are based on the use of 6 quadratic CPT strips in the half-laminate model. For the curves correspond to the infinite number of layers, there is a clear bifurcation between a primary path in which the laminate remains flat and a secondary path which follows buckling into a non-flat configuration. For a finite number of layers, the bifurcation behaviour is absent entirely due to the presence of the B_{ij} coefficients. Of course, the influence of the coupling terms is greatest for the least number of layers.

In Table 4 numerical values are given for central deflection and the force factor at two typical levels of prescribed end shortening strains well beyond the pseudo-buckling level, for laminates having 8 equal thickness layers in a $[0/90]_4$ assembly with both $A/h = 20$ and $A/h = 100$. These results are based on the use of 6 quadratic strips in the half-laminate model. The force factor is defined in equation (7.1) with the use of E_T instead of E in the equation. It can be seen that the CPT results compare closely with the SDPT results for the thin geometry but differ significantly for the thick geometry.

Figure 7 shows how longitudinal force varies with applied end shortening strain and with central deflection, as determined using the SDPT and CPT finite strips, for the eight-layer thick laminate, with $A/h = 20$.

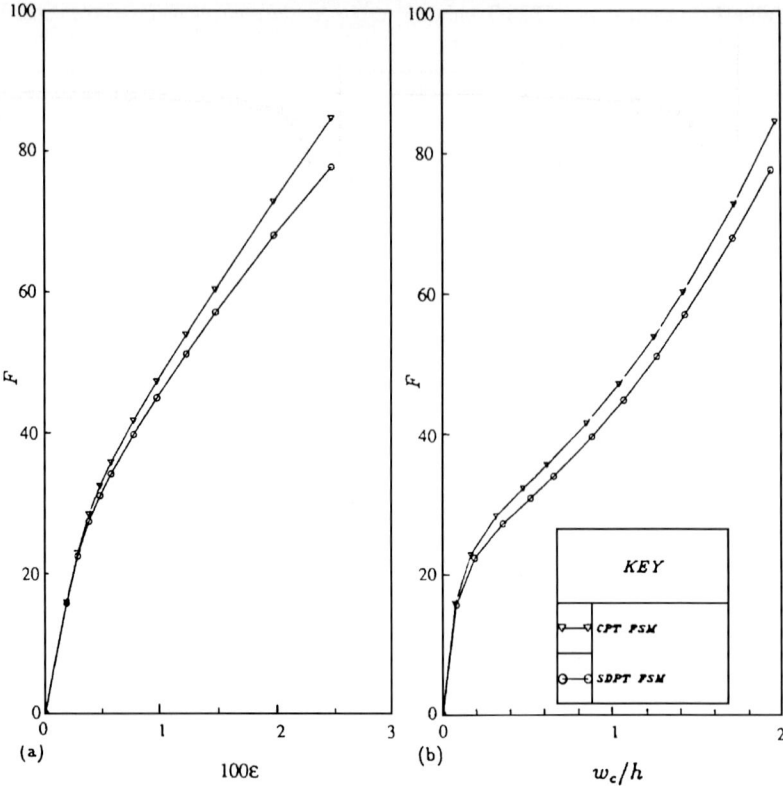

Figure 7. (a) Longitudinal force — end shortening behaviour of eight-layer cross-ply square laminates. (b) Longitudinal force — central deflection behaviour of eight-layer cross-ply square laminates.

5.7.4. Square Unsymmetric Angle-Ply Laminates

The material properties for these laminates are the same as those used in Section 7.3. The laminates are of +30/–30/+30/–30/+30/–30/... angle-ply construction with a variable (but even) number of layers. The laminates are simply supported all round for out-of-plane behaviour. In the plane of the laminates, no lateral expansion of the loaded ends is allowed and, hence, the analysis is of the Type B category and, furthermore, the unloaded edges are completely restrained against lateral expansion.

For the angle-ply laminates the non-zero coupling coefficients are B_{16} and B_{26} but, despite their presence, bifurcational buckling does occur under the action of progressive end shortening. The whole laminate is modelled with

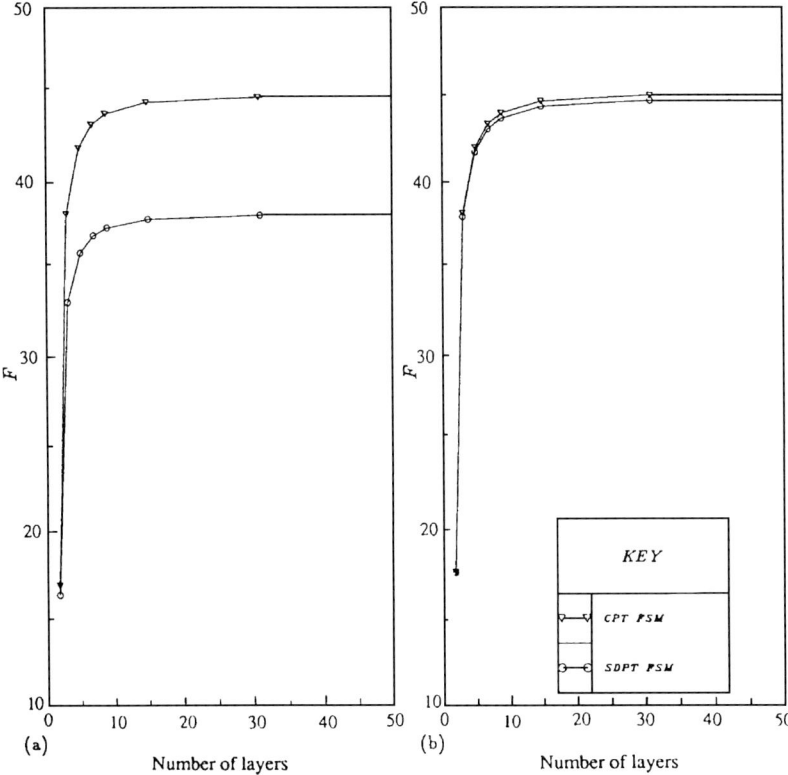

Figure 8. Variation of buckling load with number of layers for angle-ply square laminates: (a) $A/h = 20$; (b) $A/h = 100$.

12 quadratic strips and the series representation for u, v, w, ψ_x and ψ_y are sin1,2,3,4,5,6, sin1,2,3,4,5,6, sin1,2,3, cos1,2,3 and sin1,2,3 respectively. Figure 8 shows the way in which buckling load varies with the number of layers for laminates corresponding to $A/h = 20$ and $A/h = 100$, using both the SDPT and CPT finite strips. Clearly, the effect of through-thickness shear is considerable for the thicker laminate.

A comparison between SDPT and CPT predictions of the full response history, into the post-buckling range, of the eight-layer laminate with $A/h = 20$, is given in Figure 9. The effect of through-thickness shear deformation is very significant in its influence on the load levels that correspond to a given end shortening strain or a given central deflection. However, the influence is not significant on in-plane stiffness: the immediate post-buckling stiffness, in particular, is the same whichever plate theory is used.

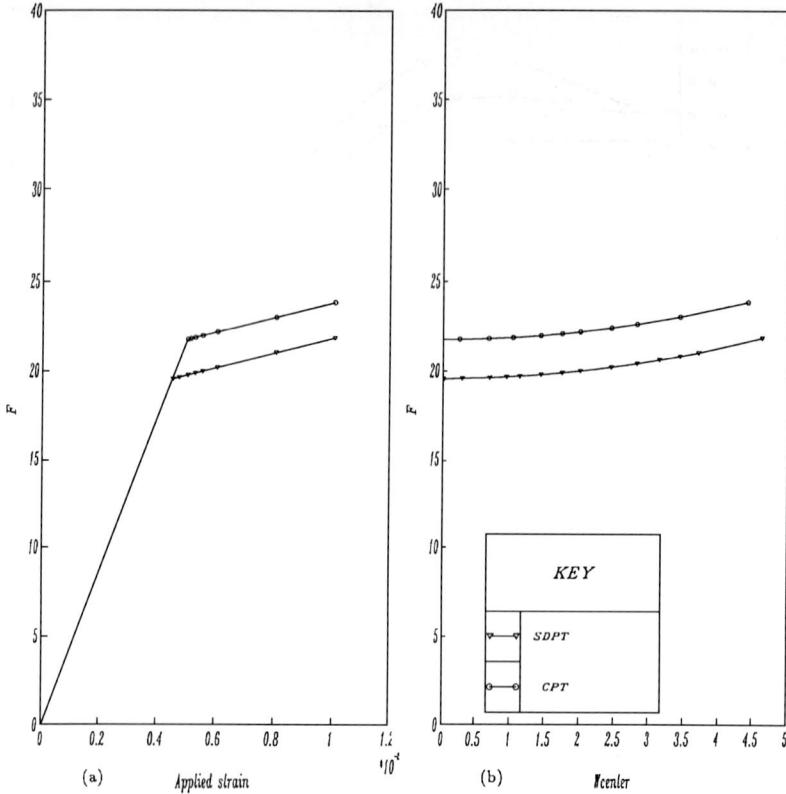

Figure 9. (a) Longitudinal force — end shortening behaviour of [+30/–30]$_4$ angle-ply square laminate. (b) Longitudinal force — central deflection behaviour of [+30/–30]$_4$ angle-ply square laminate.

All the above results are for the angle-ply laminate with fibre angle at ±30°. The manner in which the buckling stress resultant N_{xcr} and the immediate post-buckling stiffness change with the fibre angle ϕ is shown in Figure 10 for the four-ply laminate and for the laminate with an infinite number of layers (i.e. equivalent to ignoring the effect of B_{ij} terms). In Figure 10(b) the two curves coincide, i.e. the initial post-buckling stiffness is unaffected by the number of layers present. The sharp change that occur in Figure 10 at an angle of approximately $\phi = 77°$ is mainly due to a change in mode shape.

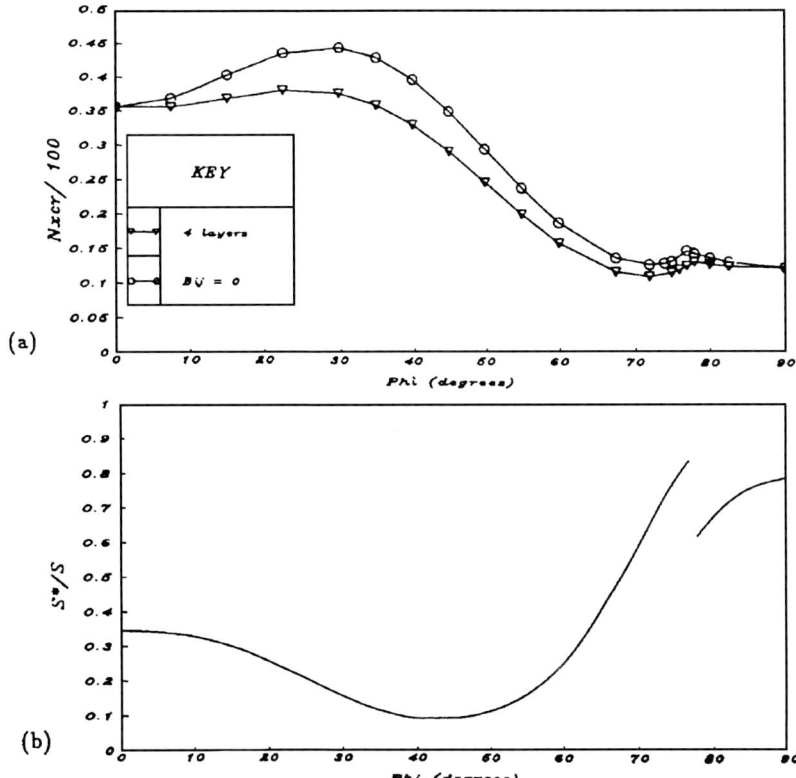

Figure 10. Variation of (a) buckling stress resultant N_{xcr}, and (b) initial post-buckling stiffness S^*/S, with respect to fibre angle f for angle-ply square laminates.

5.7.5. Rectangular 15-Layer Anisotropic Laminate

The laminate has a length of 400 mm, a width of 150 mm and a thickness of 2.25 mm. It is formed of 15 layers of equal thickness and common material. The laminate layup is [+60/–60/+30]$_5$ and the material properties are

$E_L = 104 \text{kN/mm}^2$, $E_T = 8.9 \text{ kN/mm}^2$, $G_{LT} = 5.5 \text{ kN/mm}^2$, $G_{TT} = 2.3 \text{ kN/mm}^2$, $v_{LT} = 0.32$

It follows that all the A_{ij}, D_{ij} and the B_{11}, B_{16}, B_{22} and B_{26} coefficients are non-zero, i.e. material anisotropy and coupling are present. The problem is

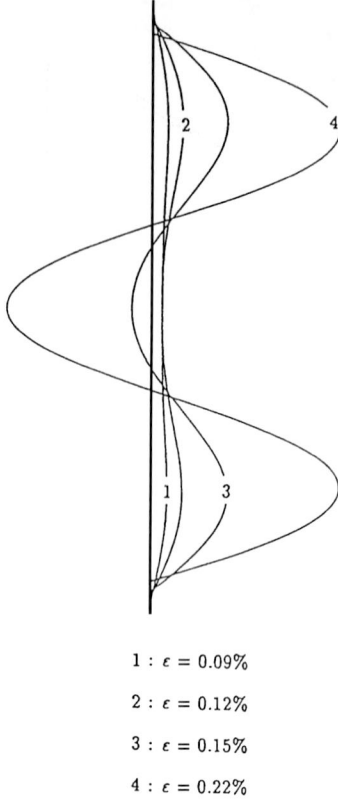

1 : $\varepsilon = 0.09\%$

2 : $\varepsilon = 0.12\%$

3 : $\varepsilon = 0.15\%$

4 : $\varepsilon = 0.22\%$

Displacements are magnified by a factor of 50.

Figure 11. Variation of w along the centre line at four levels of strain of 15-layer anisotropic rectangular laminate.

of Type B category with the loaded ends (of width 150 mm) held against in-plane lateral expansion whilst the longitudinal edges are free to expand. The laminate is simply supported all round for out-of-plane behaviour.

The whole laminate is modelled by 6 quadratic finite strips and the longitudinal series terms representation is sin1,2,...,7, sin1,2,...,7, sin1,2,...,5, cos1,2,...,5 and sin1,2,...,5. The laminate is also analysed using finite element approach[19] and a regular mesh of 4×12 elements is used over the whole laminate.

The deflection pattern of the laminate at relatively high values of end shortening strain involves three half-waves along the laminate, of slightly skewed form. The manner in which the deflected form changes at progressive increase in shortening strain is shown in Figure 11. The deflected form

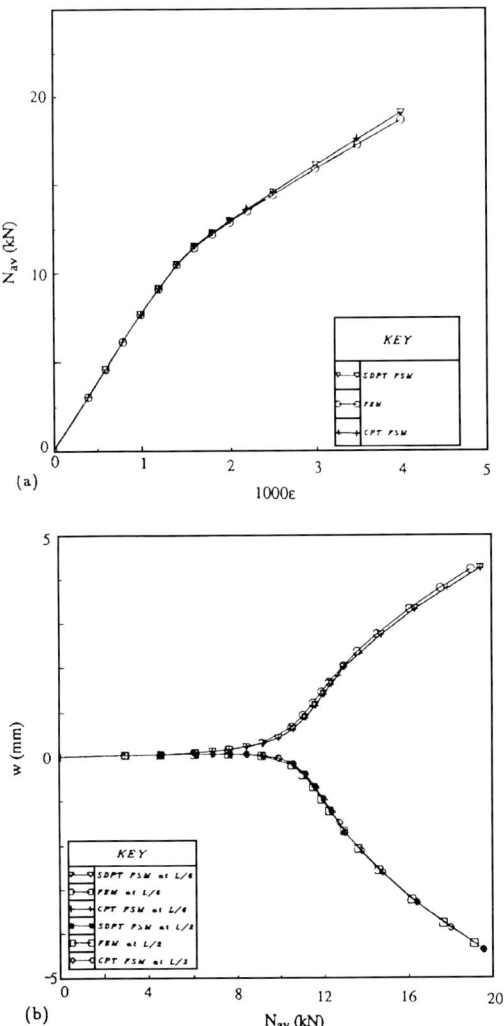

Figure 12. (a) Longitudinal force — end shortening behaviour of 15-layer anisotropic rectangular laminate. (b) Longitudinal force — central deflection behaviour of 15-layer anisotropic rectangular laminate.

at low strain levels is of markedly different shape to that at high strain levels which also implies the need to use relatively comprehensive series terms to represent the longitudinal variations of displacements.

The way in which longitudinal force and deflections, at two specific points (on the longitudinal centre line A/6 and A/2 from one end) vary with increasing end shortening is shown in Figure 12. There is no apparent difference between the SDPT and CPT results, implying that through-thickness shear effects are negligible for this relatively thin laminate. The finite element results also compare closely with the SDPT finite strip results. This also suggests that the series terms used in the representation of the longitudinal variations of the displacements are adequate.

5.8. CONCLUSION

Procedures for the prediction of the geometrically nonlinear response of rectangular laminated plates subjected to uniform end shortening have been presented based on the finite strip method within the context of SDPT and CPT. The methods have been applied the solution of a number of specific applications involving different kinds of laminates. It has been shown that markedly different forms of response can occur from one laminate to another. There is close comparison between the predictions of the SDPT finite strip and the finite element methods. This has served to verify the finite strip method.

The through-thickness shearing effect which is accounted for in the SDPT but not in the CPT has also been addressed. Clearly, the magnitude of the effect varies greatly with the problem under consideration but the effect can often be of importance in the analysis of laminates and should not be ignored in laminates of other than very thin geometry. Further to this the effect is generally more prominent to the buckling stress than to the initial post-buckling stiffness.

ACKNOWLEDGEMENT

The authors are pleased to acknowledge the financial support of the Department of Trade and Industry (through the Royal Aerospace Establishment, Farnborough) during the period of the study described here.

References

1. Middleton, D.H., 1990, "The Harrier AV-8B/GR5". In D.H. Middleton, editor, *Composite materials in aircraft structures*. John Wiley and Sons, N.Y.
2. Leissa, A.W., 1987, "A review of laminated composite plate buckling". *Appl. Mech. Review*, **40**, 575–591.
3. Kapania, R.K. and Raciti, S., 1989, "Recent advances in analysis of laminated beams and plates, Part I: shear effects and buckling". *AIAA Journal*, **27**, 923–934.
4. Reissner, E., 1945, "The effect of transverse shear deformation on the bending of elastic plates". *J. Appl. Mech.*, **12**, 69–77.
5. Mindlin, R.D., "Influence of rotary inertia and shear on flexural motion of isotropic elastic plates". *J. Appl. Mech.*, **18**, 31–38.

6. Dawe, D.J. and Craig, T.J., 1988, "Buckling and vibration of shear deformable prismatic plate structures by a complex finite strip method". *Int. J. Mech. Sci.*, **30**, 77–99.

7. Dawe, D.J. and Peshkam, V., 1989, "Buckling and vibration of finite-length composite prismatic plate structures with diaphragm ends, Part I: Finite strip formulation". *Comp. Methods Appl. Mech. Eng.*, **77**, 1–30.

8. Peshkam, V. and Dawe, D.J., 1989, "Buckling and vibration of finite-length composite prismatic plate structures with diaphragm ends, Part II: Computer programs and buckling applications". *Comp. Methods Appl. Mech. Eng.*, **77**, 227–252.

9. Dawe, D.J. and Peshkam, V., 1990, "Buckling and vibration of long plate structures by complex finite strip methods". *Int. J. Mech. Sci.*, **32**, 743–766.

10. Graves-Smith, T.R. and Sridharan, S., 1978, "A finite strip method for the post-locally-buckled analysis of plate structures". *Int. J. Mech. Sci.*, **20**, 833–842.

11. Sridharan, S. and Graves-Smith, T.R., 1981, "Post-buckling analysis with finite strips". *J. Eng. Mech. Div., ASCE*, **107**, 869–887.

12. Gierlinski, J.T. and Graves-Smith, T.R., "The geometric nonlinear analysis of thin walled structures by finite strips". *Thin-Walled Struct.*, **2**, 27–50.

13. Hancock, G.J., 1981, "Nonlinear analysis of thin-walled sections in compression". *J. Struct. Eng. Div., ASCE*, **107**, 455–471.

14. Bradford, M.A. and Hancock, G.J., 1984, "Elastic interaction of local and lateral buckling in beams". *Thin-Walled Struct.*, **2**, 1–25.

15. Azizian, Z.G. and Dawe, D.J., 1985, "Analysis of the large deflection behaviour of laminated composite plates using the finite strip method". In I.H. Marshall, editor, *Composite Structures — 3*, pp. 677–691. Elsevier Applied Science, London.

16. Dawe, D.J. and Azizian, Z.G., 1987, "Post-buckled stiffness of rectangular orthotropic composite laminates". In I.H. Marshall, editor, *Composite Structures — 4*, pp. 1.138–1.151. Elsevier Applied Science, London.

17. Dawe, D.J., Lam, S.S.E. and Azizian, Z.G., 1993, "Finite strip post-local-buckling analysis of composite prismatic plate structures". *Comp Struct.*, **48**, 1011–1023.

18. Dawe, D.J., Lam, S.S.E. and Azizian, Z.G., 1992, "Nonlinear finite strip analysis of rectangular laminates under end shortening, using classical plate theory". *IJNME*, **35**, 1087–1110.

19. Lam, S.S.E., Dawe, D.J. and Azizian, Z.G., 1993, "Nonlinear analysis of rectangular laminates under end shortening, using shear deformation plate theory". *IJNME*, **36**, 1045–1064.

20. Dawe, D.J. and Lam, S.S.E., 1992, "Analysis of the post-buckling behaviour of rectangular laminates", *Proc 33rd AIAA/ASME/ASCE/AHS/ASC Structures, Structural Dynamics and Materials Conference, Part 1*, pp. 219–229.

21. Dawe, D.J., Wang, S. and Lam, S.S.E., 1995, "Finite strip analysis of imperfect laminated plates under end shortening and normal pressure". *IJNME*, **38**, 4193–4205.

22. Cheung, Y.K., 1968, "The finite strip method in the analysis of elastic plates with two opposite simply supported ends". *Proc I.C.E.*, **40**, 1–7.

23. Cheung, Y.K., 1976, *Finite Strip Method in Structural Analysis*. Pergamon, Oxford.

24. Whitley, J.M., 1973, "Shear correction factors for orthotropic laminates under static load". *J. Appl. Mech.*, **40**, 302–304.

25. Novozhilov, V.V., 1953, *Foundation of nonlinear theory of elasticity*. Greylock, Rochester, N.Y.

26. Dawe, D.J. and Azizian, Z.G., 1986, "The performance of Mindlin plate finite strips in geometrically nonlinear analysis". *Comput Struct.*, **23**, 1–14.

27. Yamaki, N., 1959, "Post-buckling behaviour of rectangular plates will small initial curvature loaded in edge compression". *J. Appl. Mech., ASME*, **26**, 407–414.

28. Hughes, T.J.R. and Cohen, M., 1978, "The heterosis finite element for plate bending". *Comput Struct.*, **9**, 445–450.

6. OPTIMAL DESIGN OF PROCESS PARAMETERS FOR DYNAMICAL FORGING OPERATIONS

RAMANA V. GRANDHI[1] and SHYAM SANJEEV KUMAR[2]

[1] *Professor of Mechanical and Materials Engineering, Wright State University, Dayton, Ohio 45435*
[2] *Engineering Design Systems, 7961 B Washington Woods Drive, Centerville, Ohio 45459*

ABSTRACT

Forging is a thermomechanical plastic deformation process wherein the workpiece is deformed from a relatively simple geometry to a predetermined complex shape by the application of compressive forces. In metal forming processes like forging, it is necessary to control essential processing variables like strain, strain rate, and temperature, to obtain desired microstructural and service properties in the final product. The spatial and temporal distribution of these field quantities is in turn influenced by forging process parameters such as die velocity, initial die/billet temperature, and shape of the die. This work presents a methodology for designing optimal process parameters for isothermal and nonisothermal forging processes with the goal of maintaining the system variables within 'favorable' processing windows in the strain-strain rate-temperature space.

The approach integrates nonlinear rigid viscoplastic finite element simulation of the metal forming process with mathematical programming techniques. Based on information from the finite element process model, an optimization problem is formulated with the dual objective of minimizing the strain rate variance and nodal temperature variance in the deforming workpiece. Constraints are placed on the rate of material deformation at predetermined 'sensitive' regions of the workpiece, and on the temperature paths of 'critical' boundary (die-workpiece contact) nodes. The velocity of the ram and the initial die temperature adjustment parameter are selected as design variables. The objective function and constraint gradients are evaluated from the finite element system representation using a finite difference scheme. The modified method of feasible directions is employed in arriving at the optimal solution. The methodology is demonstrated with simulation case studies involving an upset forging, an axisymmetric engine disk forging, and a large-scale Integrated Blade and Rotor (IBR) disk forging.

6.1   INTRODUCTION

6.1.1.   Need for Process Control in Metal Forming

The metal forming process involves the plastic deformation of a material having a relatively simple starting geometry in one or more operations into a product of relatively complex configuration. During manufacturing the mechanical and service properties of the final product are dependent to a large extent on the prior processing history. As such, one of the most important tasks is the selection and control of critical process parameters that would ensure required part quality along with specific mechanical and physical characteristics. There is thus a strong need to develop optimal processing strategies that would result in defect-free, high quality products on a repeatable basis. Computer-aided design techniques have generally been found to be the most effective and efficient way to meet this challenge and are widely in use these days. In recent years finite element modeling and analysis has become an efficient and powerful computer aided design tool. This work endeavors to develop a process control strategy for forging operations based on finite element simulations in conjunction with numerical optimization methods. Although this work focuses on forging, the methodology featured is general purpose and is equally applicable to any of the other unit forming processes like extrusion, rolling, and sheet metal forming.

Forging is a bulk metal forming process in which a workpiece is deformed between two dies by the application of compressive forces. Open die forging is carried out between flat dies or dies having simple shapes, with no lateral constraint. In closed die forging, the workpiece is deformed to fill the die cavity representing the final component shape (Figure 1). Metal flow is restricted to fill the closed die cavity, and excess material flows through the gap between the closing dies and forms a flash around the forging at the die parting line. The flash is subsequently trimmed to form the final product (Figure 2).

Engineering components are frequently designed to meet specific functional requirements based on their application. Manufacturing a new product by forging involves selection of starting billet geometry, design of die shapes, design of number of stages required to deform the material, and determining optimal process conditions like ram velocity, temperature and friction parameters. In the past the forging die design procedure was based on the experience and intuition of the die designer and some empirical guidelines — a process entailing considerable expense and long lead times. The need for a wider variety of forgings and faster design procedures coupled with increasing costs led to the development of numerically based design techniques like the one presented in this work.

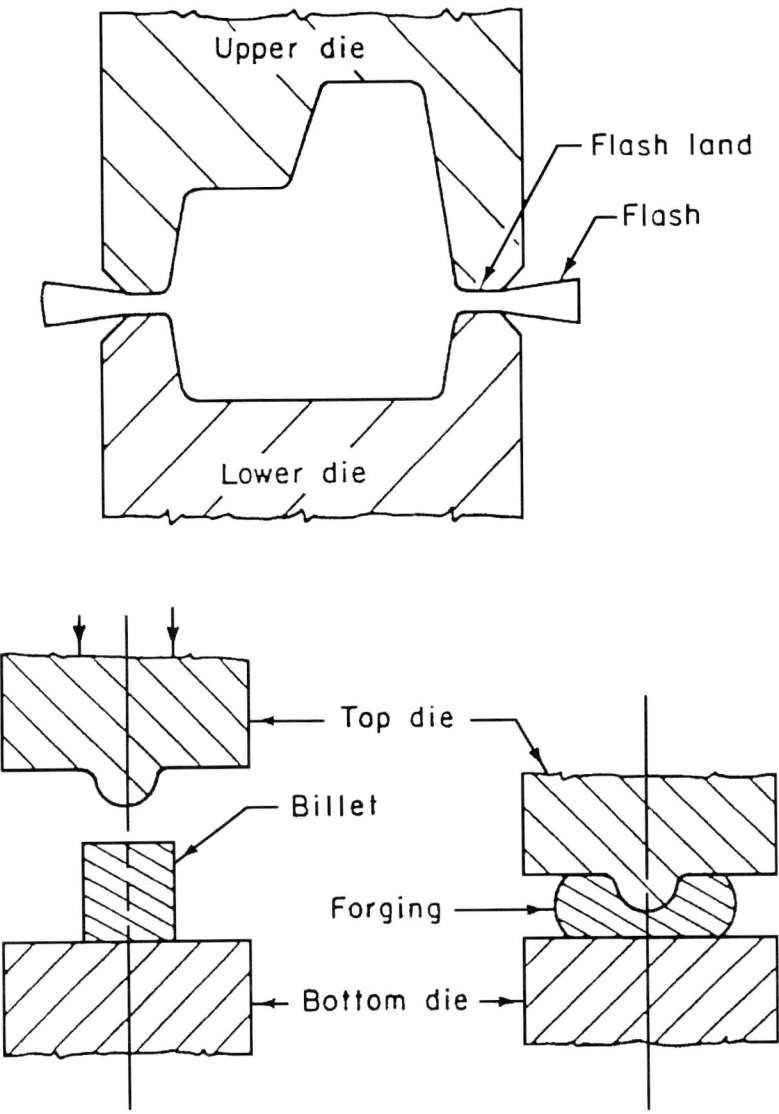

Figure 1. Closed-die and open-die forging.

Forging differs from other shaping methods in that the flow of metal is intended to produce specific material properties in the final product. The forging operation can be visualized as a system with a large number of interacting factors like starting billet shape, interface frictional conditions,

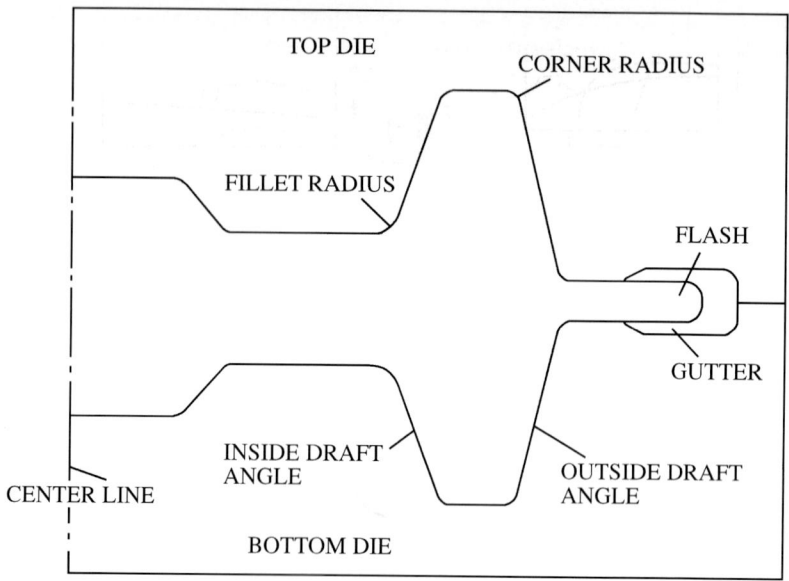

Figure 2.   Closed die forging terminology.

temperature of workpiece and dies, velocity of the die, and geometry of the
final part (Figure 4). These parameters strongly influence the thermomechanical behavior of the deforming material and have a direct impact on
the spatial and temporal distribution of field variables like strain rate, total
effective strain, and nodal temperatures. This is depicted pictorially in
Figure 3 which shows a typical forging system in which the billet is com-
pressed between two dies. The lower die is kept stationary while the upper
die is moved downwards with a velocity v to exert a compressive load
(represented by F) on the workpiece. The flow stress of the workpiece
material is given by σ_{fb}, while T_D and T_b represent the die and billet
temperatures, respectively. To reduce excessive frictional effects and improve die-life a suitable lubricant is introduced at the die-billet interface.
ε and $\dot{\varepsilon}$ represent the effective strain and effective strain rate in the deforming workpiece. The relationship between the system variables is depicted by
means of the plot on the right hand top corner of Figure 3 which shows the
variation in strain rate with flow stress for different billet temperatures at
constant strain under a given ram velocity.

The properties and integrity of the final formed product are, in turn,
functions of the field components of the metal forming system. It is thus
necessary to monitor and control the essential field variables like strain,

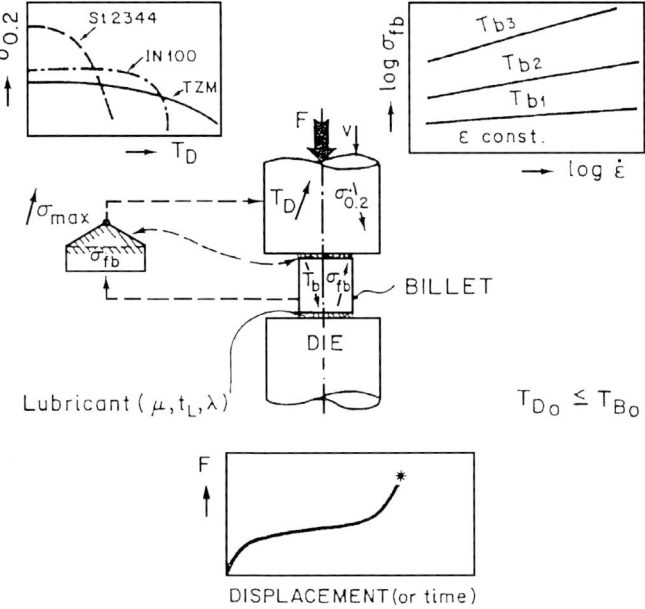

Figure 3. The forging system.

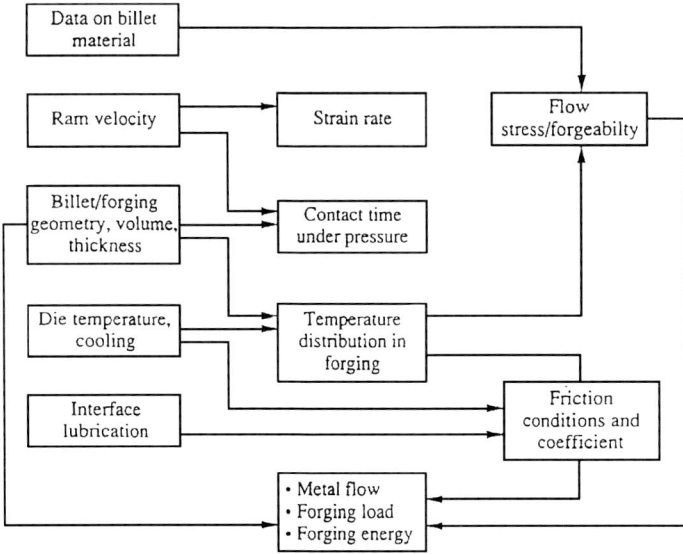

Figure 4. Interaction between forging system variables.

strain rate, and temperature to obtain the desired microstructure and service properties in the final product. Also, non-conventional and difficult-to-process materials generally have a narrow range of processing conditions in the strain, strain rate, and temperature space where they can be processed successfully without any formation of defects. To maintain the field variables within these 'favorable' processing windows, optimal design of process parameters such as die velocity and initial die/billet temperature must be carried out.

6.1.2. Process Parameters and Physical/Microstructural Properties

Generally, metal forming processes may be divided into two categories: massive forming and sheet metal forming. The massive forming processes include operations such as forging, extrusion, and rolling, while the sheet forming processes include operations like bending, spinning, and deep drawing. In these processes there is a strong relationship between macroscopic (physical) properties, microstructure, and process parameters. This is strongly supported by earlier work done in this area. For instance, in the design of aircraft engine disks, attempts have been made to obtain desired properties in the final product by generating different microstructures in different regions of the disk.[1] Though detailed relationships between microstructure and physical properties have not yet been fully established, it has been shown through experiments that physical properties depend to a large extent on the microstructure of the material, and if the workpiece is not processed in the right manner and under appropriate operating conditions, defective products and/or unwanted physical characteristics in the final product may result.

In metal forming processes, it has generally been found that process parameters play an important role in the forming of a specific microstructure in the workpiece. For example, Devadas et al.,[2] modeled the microstructure and mechanical properties of steel during hot rolling. These models have been used to predict grain size and characterize static and dynamic recrystallization during the rolling process. Most of these models have been found to be sensitive to change in process parameters such as strain, strain rate, and temperature. Sastry et al.[3] studied the relationship between metallurgical properties and process parameters during the superplastic forming of titanium alloys. This work showed that the percentage of equilibrium α phase in the microstructure is related to strain rate, strain rate sensitivity, and temperature. Dadras and Thomas[4] conducted experiments to study the deformation behavior of Ti-6242 (Ti-6Al-2Sn-4Zr-2Mo-0.1Si) alloy during the upset forging of specimens starting with $(\alpha + \beta)$ or β microstructure. They found that the volume percentage of primary α microstructure in the specimen is influenced by the deformation temperature. This further illustrated and emphasized the importance of temperature effects during deformation. Cohen and Durham[5] correlated the effect of change in strain rate to the final microstructure during hot working processes. The dependence of the resulting microstructure on various temperatures and strain rates was analyzed while carrying out compression tests on a carbon steel material. Seetharaman *et al.*[6] analyzed the effect of strain, strain rate, and temperature on the microstructure of a Gamma Ti-Al alloy during extrusion. During this work, again, a definite relationship between microstructural parameters (grain size and grain distribution) and process variables (effective strain rate, effective strain, and flow stress) was observed. This further proved that both strain rate and temperature have a strong influence on the microstructure and properties of a given material.

The following major parameters have a strong influence on the forging process:

• Ram speed
• Initial die/workpiece temperatures
• Lubricant
• Preform geometry and staging criteria
• Die/workpiece material
• Type and size of the forging press

In the past, the design of these process parameters was done on a trial and error basis, and was dependent to a large extent on the experience of the die designers, metallurgists and process engineers. This method was expensive, time consuming, and tedious. With the development of sophisticated numerical techniques, computer aided design and analysis methods are steadily gaining in popularity.[24]

6.1.3. Metal Forming Simulation Using the Finite Element Method (FEM)

In the past, a number of approximate methods of analysis have been developed and applied to various forming processes. Some of the well known metal forming simulation and analysis methods are the slab method, the upper-bound method, the slip-line field method, the visioplasticity method, and the finite element analysis (FEA) method.

In the slab method, the workpiece being deformed is decomposed into several slabs. For each slab, a simplified stress distribution is assumed. The resulting approximate equilibrium equations are solved with imposition of stress compatibility between slabs and boundary tractions. Although this method is quick and gives reasonable results,[7,8] its main shortcoming is that it is restricted to the evaluation of loads and stresses for simple geometries only.

The upper bound method[9,10] requires the guessing of admissible velocity fields, among which the best one is chosen by minimizing the total potential energy. Based on this velocity field, the total forming energy and the forming load are then calculated. This approach has earlier been used in the analysis of axisymmetric forgings, plate rolling, and extrusion processes.[11–15]

The visioplasticity method[16] combines experiment and analysis. After the velocity vectors have been determined from an actual test, strain rates are calculated, and the stress distributions are obtained from plasticity equations.

All of the above methods have been useful in predicting forming loads, overall geometry changes of the deforming workpieces, qualitative modes of metal flow, and in determining optimum process conditions. However, accurate determination of the effects of various process parameters on the detailed metal flow became possible only with the development of the finite element method (FEM). The appeal of the finite element method stems from its ability to systematically represent material behavior and the complex boundary conditions associated with metal forming processes. The finite element method also has the capability of producing detailed deformation information under combined mechanical and thermal loading, for solving a large variety of problems.

In recent years, due to the rapid development of computers and numerical algorithms, FEM has become very popular for simulating and analyzing metal forming problems.[24] The FE methods are generally based on the flow formulation or solid formulation. For rigid-plastic materials, it is assumed that flow stress is a function of strain, strain rate, and temperature, and that the elastic response of the material is negligible. The rigid-viscoplastic theory was generalized by Zienkiewicz et al.[17,18] and is capable of modeling hot, rate-dependent processes. This generalization provides the theory for analyzing the deformation of Ti-alloys which are strain rate sensitive materials. In other applications, phenomena associated with elasticity cannot be neglected. In the so-called solid formulation,[24] the material is considered to behave as an elastic-plastic or elastic-viscoplastic solid.

Lee and Kobayashi[19,20] developed a "general matrix method" based on the variational principle for rigid-plastic deformation. The method proved to be very well suited to large deformation plasticity problems and was applied to solid cylinder compression, plane stress bore expansion and flange drawing. In the early 80's, Oh et al.[21,22] developed the rigid-viscoplastic formulation to solve a wide variety of problems using the finite element method. These efforts resulted in the development of a generalized computer program ALPID[23,24] (Analysis of Large Plastic Incremental Deformation), which has the capability to perform a wide range of 2-D metal forming simulations and analyses.

Besides the rigid-plastic and rigid-viscoplastic finite element analysis, some researchers also used other plastic theories. Dexter[25] studied the mechanisms involved in forging using the elastic-viscoplastic finite element analysis. The stress and strain values, especially at the die-workpiece interface, and the critical loads for the potential failure of forging die can be obtained by this method of analysis. The updated Lagrangian Jaumann Formulation of FEM, which includes elastic deformation and tends to be elastic-plastic or elastic-viscoplastic in nature, has also been used in metal forming.[26]

Wu and Oh[27,28] further incorporated a thermal module into the ALPID program for performing thermo-viscoplastic deformation analyses. Zienkiewicz et al.[29] studied and performed a coupled thermal analysis for steady-state extrusion operations. Tang et al.[30] analyzed the shell nosing problem using this approach. Coupled analysis using the updated Lagrangian approach also has been reported in earlier work.[31] More recently, a general purpose 3-D analysis program, ANTARES (A Natural Tool to Acquire the Required Engineering Shape), with automatic remeshing has been developed[32] for metal forming analysis and simulation.

In this work, the finite element method has been chosen as the primary numerical analysis tool. The nonlinear rigid viscoplastic finite element program ALPID has been used for simulation and analysis purposes because it has the following capabilities:

1. Obtaining detailed solutions of mechanics in a deforming body with sufficient accuracy for practical purposes. The solution contains velocities, strains, stresses, strain rates, temperatures, contact pressure distributions, die load and die temperatures.
2. Handling arbitrary boundary conditions.
3. Including the friction effect at the die-workpiece interface in both deformation and thermal calculations.
4. Graphics display for post-processing.
5. Analyzing a large variety of problems by simply changing the input data.

6.1.4. Design of Optimal Process Parameters

Compared to traditional design methods in the metal forming field, numerical analysis and design techniques are less time consuming and expensive. Boer et al.[33] developed a process model based on the slab method to calculate stress and strain distributions in the deforming workpiece. Thermal analysis was performed using finite element methods and a simplified non-dimensional analysis. Using the above strategy, and minimizing the stress ratio parameter, optimal ram velocity profiles and initial die temperature were obtained for both isothermal and hot die forgings. Lanka, et al.[34] developed the Conformal Mapping Method to design intermediate die shapes for 2-D and 3-D forgings. This is a geometry mapping technique wherein the staging criteria are identified based on the stress ratio parameter. Malas[35] developed an approach for process parameter design using a linear relationship between the ram velocity and strain rate as an approximation to maintain the billet variables within stable processing regions. Hong[36] designed a control scheme based on a finite element analysis model. This model utilizes the local thickness of the deforming blank as a measured variable to track a desired trajectory. Park et al.[37] developed the "backward tracing" technique which is based on the reversal of material flow during simulation. Using this method preforms were designed for shell nosing, plane strain rolling, and axisymmetric forging problems. Han et al.[38] and Srinivasan et al.[39] combined this idea with a numerical optimization approach, and designed optimal intermediate die shapes for isothermal forging processes by minimizing the strain rate variance in the deforming workpiece. This technique is called the "Backward Deformation Optimization Method", and includes sensitivity analysis, besides introducing a criterion for nodal separation from the surface of the die during backward deformation simulation. Grandhi et al.[40,41] then introduced an optimal control design algorithm into the process parameter design procedure. The metal forming process was modeled and condensed into the state space form, and a suitable optimal control algorithm was used in designing the process parameters. Optimum ram velocity schedules for maintaining specified strain rates in the billet were generated by this approach for an isothermal disk forging.

During conventional hot die forging, there is a complex thermal interplay between the workpiece, die(s) and the atmosphere. There is heat generation due to the expenditure of deformation energy and due to friction. At the same time there is heat loss from the system due to conduction, convection, and radiation. As a result, the processed material may experience severe temperature gradients along the interface of the die and billet, which could lead to the formation of surface defects. Thermal disparities and temperature changes in the workpiece may also induce phase transformations and changes in the grain structure. These changes, in turn, affect the flow stress and metal flow as well as other process variables. Furthermore, severe temperature gradients result in large thermal stresses leading to material failure.

In view of these temperature induced effects, Cheng, et al.[42] extended the isothermal study explained in reference 40 to include and handle nonisothermal situations like hot die forging. The coupled deformation and thermal analysis code, ALPID,[43] was modified to build a thermo-mechanical state space model, and a finite element based condensation scheme was used to reduce the number of states involved. An optimal finite time controller design algorithm was integrated with the ALPID code to determine the required (optimal) process parameters. The results show that the above methodology was quite successful in satisfying the required design objectives. But developing the state space model from the finite element equations and solving the resulting set of equations is a non-trivial task, especially while dealing with large scale systems. While simulating realistic manufacturing processes, it is necessary to use large-scale finite element models with a large number of degrees of freedom. In such instances, the corresponding state space model would also have a large number of states, resulting in numerical difficulties during implementation, besides being tedious and computationally expensive. In such situations, model reduction techniques have to be used to reduce the order of the system before designing the controller. Grandhi *et al.*[44] developed reduced order analytical models for representing metal forming systems using state space theory. Sophisticated model reduction methods were used to build reduced order models that retained the behavioral and response characteristics of the original full order system. Using this approach optimal die velocity schedules were developed for various isothermal and nonisothermal forging examples. The above methods based on optimal control theory have their limitations when dealing with large scale finite element systems. In such cases they tend to be unjustifiably tedious and expensive to implement. It was also determined that multi-element strain rate control was not feasible using this approach. There were also restrictions on the number and nature of the state variables that could be chosen during the design process. There was thus a need to develop a relatively simple and generic technique for designing optimal process parameters for metal forming systems.

This work presents a methodology for optimal process parameter design based on numerical modeling and simulation of metal forming processes in conjunction with dynamic mathematical optimization techniques. The nonlinear rigid viscoplastic finite element method is used for simulation and analysis of the forging process. An optimization based design/control module has been developed and interfaced with the finite element system to set up and solve any specified optimization problem(s) related to the metal forming process.

The objective of this work was to develop a methodology to design optimal process parameters (ram velocity, initial die temperature, and die fillet radius) that satisfies some or all of the following requirements:

1. Maintain the strain rate at a certain value at a given region (element) in the billet.
2. Maintain the temperature at a certain value at specified critical locations (nodes) in the workpiece.
3. Reduce the temperature range in the billet.
4. Reduce the temperature gradient at the die-workpiece interface.
5. Force the strain rate and temperature (at pre-specified locations) in the billet to follow desired trajectories.
6. Obtain uniform deformation characteristics by minimizing the strain rate variance and/or temperature variance in the deforming workpiece.

The above requirements have significant practical meaning, as explained in later sections of this work. If these requirements are satisfied, the quality of the forged product can be significantly improved besides reducing the manufacturing and development cost.

6.2. SYSTEM MODEL FOR DEFORMATION ANALYSIS

6.2.1. Basis for the Finite Element Formulation

The finite element method (FEM) used in this work is based on the theory of viscoplasticity. This approach assumes that during metal forming the plastic strains significantly exceed the elastic strains and the idealization of rigid-plastic or rigid-viscoplastic material behavior is acceptable. The formulation is briefly described below:[24]

Any metal forming problem is basically a solid mechanics boundary value problem in which we have a body (workpiece) of volume V bounded by a surface S. Compatibility and equilibrium equations apply within the volume. The boundary surface S may be divided into three distinct parts given by,

$$S = S_u + S_F + S_c \qquad (2.1)$$

where, S_u is the surface with prescribed velocity $\tilde{u}$, S_F is the portion of the workpiece with traction F prescribed, and S_c is the tool-workpiece interface surface where the frictional stress f_s acts. The body is composed of a rigid plastic material which obeys the von Mises yield criterion and its associated flow rule. The deformation of the body is now characterized by the following field equations.

Equilibrium conditions for deformation analysis (neglecting body forces) in the tensor form are given by,

$$\sigma_{ij,j} = 0 \qquad (2.2)$$

where, σ_{ij} is the ij^{th} component of the Cauchy stress tensor. In the above equation, a recurring letter suffix indicates the sum, and the comma denotes partial differentiation. Compatibility conditions are given by the following strain rate-velocity relation,

$$\dot{\varepsilon}_{ij} = \frac{1}{2}(u_{i,j} + u_{j,i}) \tag{2.3}$$

where, $\dot{\varepsilon}_{ij}$ is the ij^{th} component of the strain rate tensor and u_i is a velocity component. The constitutive equation giving the stress-strain rate relationship is:[24]

$$\sigma'_{ij} = \frac{2}{3}\frac{\overline{\sigma}}{\dot{\overline{\varepsilon}}}\dot{\varepsilon}_{ij} \tag{2.4}$$

where σ'_{ij} is a component of the deviatoric stress, $\overline{\sigma}$ and $\dot{\overline{\varepsilon}}$ are the effective stress and effective strain rates, respectively, defined by $\overline{\sigma} = \sqrt{\frac{3}{2}\sigma_{ij}'\sigma_{ij}'}$ and $\dot{\overline{\varepsilon}} = \sqrt{\frac{2}{3}\dot{\varepsilon}_{ij}\dot{\varepsilon}_{ij}}$. The flow stress, in general, is a function of total strain, strain rate, temperature, and microstructure M, and may be represented as:

$$\overline{\sigma} = f(T, \overline{\varepsilon}, \dot{\overline{\varepsilon}}, M) \tag{2.5}$$

The boundary condition for this system representing equilibrium of stresses is given by,

$$\sigma_{ij}n_i = F_j, \quad \text{on} \quad S_F \tag{2.6}$$

where, n_i is the ith-component of the unit normal vector to the body surface and F_j is the jth-component of the prescribed surface traction. The velocity boundary condition is given by,

$$u_i = \tilde{u}_i, \quad \text{on} \quad S_u \tag{2.7}$$

where $\tilde{u}_i$ is the ith-component of prescribed velocity.

Solutions to this problem are the stress and velocity distributions that satisfy the governing equations and the boundary conditions. The governing equations include the equilibrium equations, the yield criterion, and the compatibility conditions derived from the flow rule. Since it is difficult to obtain a complete solution that satisfies all the field equations, various approximate methods are used to solve the above problem, one of them being the finite element method. The basic principles and concepts involved in the finite element method are the variational principle and discretization.

The finite element governing equation for metal forming analysis is derived from the potential energy functional,[24]

$$\pi = \int_V \bar{\sigma}\dot{\bar{\varepsilon}}\,dV - \int_{S_F} F_i u_i\,dS \tag{2.8}$$

where the first term represents the rate of total plastic work done and the second term represents the rate of work expended due to surface traction. The variational principle requires that among admissible velocities u_i that satisfy the conditions of compatibility and incompressibility, as well as the velocity boundary conditions, the actual solution gives the above functional a stationary value. Therefore, at equilibrium,

$$\delta\pi = \int_V \bar{\sigma}\delta\dot{\bar{\varepsilon}}\,dV - \int_{S_F} F_i \delta u_i\,dS = 0 \tag{2.9}$$

Metal deformation occurs at a constant volume, and the admissible velocity u_i needs to satisfy the incompressibility constraint. This is embedded into the variational principle functional (Eq. (2.9)) by introducing a penalty term Q (which is a very large positive quantity) into the equation as,

$$\delta\pi = \int_V \bar{\sigma}\delta\dot{\bar{\varepsilon}}\,dV + Q\int_V \dot{\varepsilon}_V \delta\dot{\varepsilon}_V\,dV - \int_{S_F} F_i \delta u_i\,dS = 0 \tag{2.10}$$

where $\dot{\varepsilon}_V = \dot{\varepsilon}_{ii}$ is the volumetric strain rate. The development of the finite element equations from the above equation is described in the next section. The traction boundary condition on S_F is either zero-traction or ordinarily a uniform hydrostatic pressure. However, the boundary conditions at the die-workpiece interface (S_C) are mixed. In this region, the velocity boundary condition is given in the direction normal to the interface by the die velocity, and the traction boundary condition is expressed by,

$$f_s = mkl = mk\frac{2}{\pi}\tan^{-1}\frac{|u_s|}{u_0}l \tag{2.11}$$

where, f_s is the frictional stress, l is the unit vector in the direction opposite to the sliding, u_s is the sliding velocity of a material relative to the die velocity, and u_0 is a small positive number compared to u_s. m is the friction factor for the constant shear force friction model, and k is the shear yield stress.

6.2.2. Finite Element Equations

The discretization of the functional described above is done using the standard procedure of the finite element method. The primary unknown for the solution of Eq. (2.10) (which represents a quasi-static plastic deformation process) is the velocity field associated with it. This velocity field, u, is approximated by shape functions in terms of nodal point velocity values as,

$$u = \mathbf{N}_D^T \mathbf{v} \qquad (2.12)$$

where, $\mathbf{N}_D$ is the shape function matrix for deformation analysis, and $\mathbf{v}$ is the nodal velocity vector for the element under consideration. Eq. (2.10) may now be expressed in terms of the nodal point velocities $\mathbf{v}$ and their variations Using the variational principle, from the arbitrariness δv_I of a set of nonlinear algebraic equations (stiffness equations) are obtained as shown below:[24]

$$\frac{\partial \pi}{\partial v_I} = \sum_j \left(\frac{\partial \pi}{\partial v_I} \right)_{(j)} = 0 \qquad (2.13)$$

where, (j) represents the quantity at the jth element. The capital-letter suffix, I, refers to the nodal point number. The above equation is actually determined by first obtaining the elemental equations and then assembling them under appropriate constraints to obtain a set of global stiffness equations in the standard finite element form,

$$^n\mathbf{K}^n\mathbf{V} = {}^n\mathbf{F} \qquad (2.14)$$

where, $\mathbf{K}$ is the global stiffness matrix, $\mathbf{F}$ is the load vector, $\mathbf{V}$ is the global velocity vector, and n is the counter representing the iteration number at every time step. This problem is similar to a standard finite element analysis problem, but its special feature is that the geometry of the boundary, and hence the boundary condition, keeps changing as the die stroke increases. $\mathbf{K}$ and $\mathbf{F}$ are both implicit functions of $\mathbf{V}$, resulting in a highly nonlinear system of equations. Therefore, the analysis path of the forging process is traced in an incremental manner by considering a series of discrete equilibrium states, each corresponding to a specific value of time. The nth discrete equilibrium state is assumed to correspond to the time value t, and the subsequent or $(n + 1)^{th}$ equilibrium state, is assumed to correspond to time $(t + \Delta t)$. The geometry configuration at time $(t + \Delta t)$ is updated from time (t). This is known *a priori* from the solution of the previous time step (t). The velocity distribution for configuration $(t + \Delta t)$ is unknown and calculated in an implicit iterative manner until a converged solution is obtained.

The solution to Eq. (2.13) or Eq. (2.14) is generally obtained iteratively using the Newton-Raphson method or some quasi-Newton method. The procedure consists of linearization and application of convergence criteria to obtain the final solution. If the converged point is $\mathbf{v} = \mathbf{v}_0$, a Taylor series expansion can be made about $\mathbf{v}_0$ as follows:

$$\left[\frac{\partial \pi}{\partial v_I}\right]_{\mathbf{v}=\mathbf{v}_0} + \left[\frac{\partial^2 \pi}{\partial v_I \partial v_J}\right]_{\mathbf{v}=\mathbf{v}_0} \Delta v_J = 0 \qquad (2.15)$$

where Δv_J is the first order correction of the velocity $\mathbf{v}$. The linearized system of equations may be written as,

$$K_S \mathbf{v} + K_t \Delta \mathbf{v} = \mathbf{F} \qquad (2.16)$$

where, $\mathbf{K}_S$ and $\mathbf{K}_t$ are called the secant stiffness and tangent (or gradient) stiffness matrices, respectively. $\mathbf{K}_S$ and $\mathbf{K}_t$ are calculated based on the current solution of Eq. (2.14) and treated as constant matrices for time t to $(t + \Delta t)$. Here again, $\mathbf{F}$ is the finite element load vector. Once the solution of Eq. (2.15) for the velocity correction term ($\Delta \mathbf{v}$) is obtained, the current velocity $\mathbf{v}_0$ is updated as ($\mathbf{v}_0 + \alpha \Delta \mathbf{v}$), where α is a constant between 0 and 1, and is called the deceleration coefficient. Iteration is continued until the velocity correction terms become negligible and the convergence criteria are met.

To calculate the first and second derivatives in Eq. (2.15), the effective strain rate ($\dot{\bar{\varepsilon}}$) and volumetric strain rate ($\dot{\varepsilon}_V$) must be expressed in terms of the nodal velocities. This is done by means of the strain-rate matrix $\mathbf{B}_s$ defined by,[24]

$$\dot{\varepsilon} = \mathbf{B}_s \mathbf{v} \qquad (2.17)$$

where $\dot{\varepsilon}$ is the strain rate vector which contains the normal strain rate and engineering shear strain rate components. The effective strain rate, in turn, can be expressed in terms of the strain-rate vector as,[24]

$$(\dot{\bar{\varepsilon}})^2 = \dot{\varepsilon}^T \mathbf{D} \dot{\varepsilon}$$
$$= \mathbf{v}^T \mathbf{B}_s^T \mathbf{D} \mathbf{B}_s \mathbf{v} \qquad (2.18)$$
$$= \mathbf{v}^T \mathbf{P} \mathbf{v}$$

where $\mathbf{P} = \mathbf{B}_s^T \mathbf{D} \mathbf{B}_s$. $\mathbf{D}$ is a diagonal matrix that has $\frac{2}{3}$ and $\frac{1}{3}$ as its components, corresponding to the normal and shear strain rates, respectively. Also, the volumetric strain rate is defined as,

$$\dot{\varepsilon}_V = C_i v_i \qquad (2.19)$$

where, $C_i = B_{1i} + B_{2i} + B_{3i}$, and B_{ij} is an element of the strain-rate matrix $\mathbf{B}_s$. Using the above formulae, the first and second order derivatives in Eq. (2.15) may be expressed as,[24]

$$\frac{\partial \pi}{\partial v_i} = \int_V \frac{\overline{\sigma}}{\dot{\overline{\varepsilon}}} P_{ij} v_j dV + \int_V Q C_j v_j C_i dV - \int_{S_F} F_j N_{ji} dS \qquad (2.20)$$

and,

$$\frac{\partial^2 \pi}{\partial v_i v_j} = \int_V \frac{\overline{\sigma}}{\dot{\overline{\varepsilon}}} P_{ij} dV + \int_V \left(\frac{1}{\dot{\overline{\varepsilon}}} \frac{\partial \overline{\sigma}}{\partial \dot{\overline{\varepsilon}}} - \frac{\overline{\sigma}}{\dot{\overline{\varepsilon}}^2} \right) \frac{1}{\dot{\overline{\varepsilon}}} P_{ik} v_k v_m P_{mj} dV + \int_V Q C_j C_i dV \qquad (2.21)$$

The detailed derivation of the above equations and explanation of the terms involved may be obtained from reference 24.

6.3. SYSTEM MODEL FOR THERMAL ANALYSIS

6.3.1. Finite Element Equations for Heat Transfer

During metal forming operations a large amount of the mechanical work is converted into heat through large plastic deformations and through friction forces at the tool (die)-workpiece interface. The temperature distribution in the workpiece and dies in general is an important aspect to be considered during the design process because temperature affects the microstructure, which in turn influences the mechanical and service properties of the product being formed. By coupling the earlier discussed deformation model with the thermal analysis procedure, a more realistic simulation of the metal forming process may be obtained.

Therefore, in addition to deformation analysis, nonisothermal metal forming simulation also needs a comprehensive thermal analysis to determine the temperature distribution in the billet and die domains. The thermal analysis involves several separate bodies like the die, workpiece, and lubricant, and takes into account the thermal interplay between each of them. The die and billet are discretized (Figure 5), and finite element analysis for each of these bodies is conducted separately. Heat transfer between the distinct bodies through the region of contact is modeled by enforcing appropriate heat transfer boundary conditions. This section gives an overview of the thermal analysis procedure for metal forming simulation using the finite element method.[24,27]

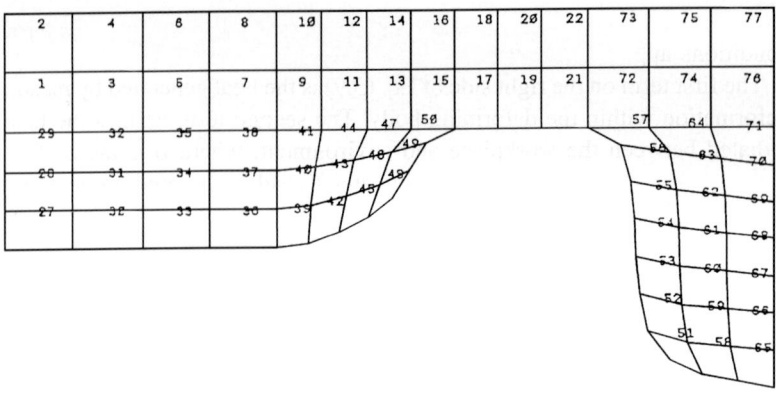

Figure 5. FEM discretization of the die body.

Consider a body in thermal equilibrium under a specific set of prescribed thermal boundary conditions (Figure 6). The energy balance equation for heat transfer analysis is given by,

$$\nabla \cdot \mathbf{q} - \dot{r} + \rho c \frac{dT}{dt} = 0 \qquad (3.1)$$

where ρ and c are the mass density and specific heat, respectively; $\dot{r}$ is the heat generation rate, T is the current temperature, and t refers to time. Also, $\mathbf{q}$ is the heat flux, and can be written using Fourier's law of heat transfer as,

$$\mathbf{q} = -\mathbf{k_c} \cdot \nabla T \qquad (3.2)$$

where $\mathbf{k_c}$ denotes thermal conductivity, and ∇T is the spatial gradient of temperature. In Eq. (3.1), $\nabla \cdot \mathbf{q}$ refers to the divergence of heat flux, which is given by, $\nabla \cdot \mathbf{q} = \frac{\partial q_m}{\partial X_m}$, where X_m is the m-coordinate value for a generic point P in the body domain. If the heat generation in the deforming body is assumed to be due to plastic deformation only, then,

$$\dot{r} = \kappa \sigma_{ij} \dot{\varepsilon}_{ij} \qquad (3.3)$$

where the heat generation efficiency κ represents the fraction of mechanical energy transformed into heat (which is generally assumed to be 0.9). In this

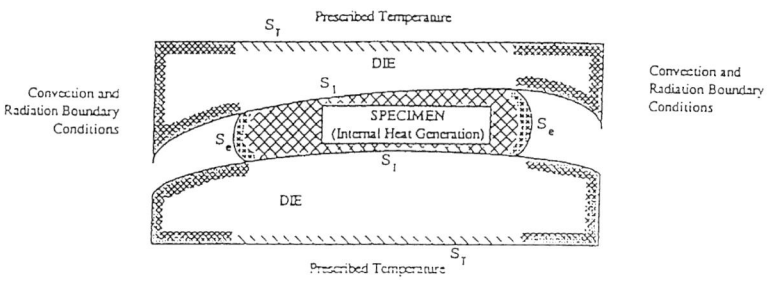

Figure 6.   Thermal boundary conditions.

work, two types of boundary conditions are considered over the body surface (Figure 6). On surface S_T, the temperature is prescribed as:

$$T = \tilde{T} \qquad \text{on} \quad S_T \qquad (3.4)$$

On surface S_q, the heat flux is specified as,

$$\mathbf{n} \cdot \mathbf{q} + \tilde{q} = 0 \qquad \text{on} \quad S_q = S_I + S_e \qquad (3.5)$$

where, $\mathbf{q}$ is the heat flux due to conduction, and $\mathbf{n}$ is a unit vector acting normal to the body surface in an outwards direction. $\tilde{q}$ represents heat flux due to convection (and/or radiation) boundary conditions and heat generated due to friction in the die-workpiece contact area. S_I is the die-workpiece interface, and S_e represents the surface exposed to the environment.

Using the Galerkin weighted residual approach, with δT (i.e., virtual temperature increment) as the weighting function, a weak form of Eq. (3.1) is obtained as,[24]

$$\int_V \left[\nabla \cdot (\mathbf{k_c} \cdot \nabla T) + \dot{r} - \rho c \frac{dT}{dt} \right] \delta T dV + \int_S (\tilde{q} + \mathbf{n} \cdot \mathbf{q}) \delta T dS = 0 \qquad (3.6)$$

where S is the total surface area of the body ($S = S_T + S_q$), and V is the current volume of the body. Solutions to problems of this nature require that the temperature field satisfies the prescribed boundary conditions and Eq. (3.6) for an arbitrary perturbation δt.

To implement the finite element procedure, the temperature field in Eq. (3.6) is discretized and approximated as,

$$T = \mathbf{N}_T^T \mathbf{T} \qquad (3.7)$$

where $\mathbf{N}_T$ is the shape function vector for thermal analysis, and $\mathbf{T}$ is the vector of nodal temperatures. By substituting Eq. (3.7) into Eq. (3.6) and expanding, the finite element governing equation for heat transfer analysis is obtained, as shown below:

$$\mathbf{C}_T \dot{\mathbf{T}} + \mathbf{K}_c \mathbf{T} = \mathbf{Q}_T \qquad (3.8)$$

where $\mathbf{C}_T$ is the heat capacity matrix, $\mathbf{K}_c$ is the heat conduction matrix, $\mathbf{Q}_T$ is the heat flux vector, and $\mathbf{T}$ is the vector of nodal point temperatures.

The transient thermal behavior of the body is captured in an incremental manner by considering a series of discrete thermal equilibrium states, each corresponding to a specific value of time. The nth discrete equilibrium state is assumed to correspond to the time value (t) and the subsequent, or $(n + 1)^{th}$ equilibrium state is assumed to correspond to time $(t + \Delta t)$ where Δt is the (current) time increment value. Therefore, for a particular $\mathbf{C}_T$ and $\mathbf{K}_c$, Eq. (3.8) is used to obtain the temperature solution in the interval $(t, t + \Delta t)$. Then, $\mathbf{C}_T$ and $\mathbf{K}_c$ are updated, and the procedure is repeated for the next time step.

In Eq. (3.8), the heat flux vector $\mathbf{Q}_T$ has several components and is expressed as,

$$\mathbf{Q}_T = \int_V \kappa(\overline{\sigma}\dot{\overline{\varepsilon}}) \mathbf{N}_T dV + \int_{S_e} \sigma\epsilon(T_e^4 - T_s^4) \mathbf{N}_T dS$$
$$+ \int_{S_e} h(T_e - T_s) \mathbf{N}_T dS + \int_{S_I} h_{lub}(T_d - T_w) \mathbf{N}_T dS + \int_{S_I} q_f \mathbf{N}_T dS \qquad (3.9)$$

where, S_e refers to the surface where convection and radiation boundary conditions are specified. S_I is the surface where interface heat transfer conditions apply.

The first term on the right side of Eq. (3.9) is the heat generated by plastic deformation within the deforming body. The second term defines the heat radiated between the workpiece and environment, where σ is the Stefan-Boltzman constant, ϵ is the emissivity, and T_e and T_s are environment and surface temperatures, respectively. The third term describes the heat convected from the body to the environment with heat convection coefficient h. The fourth term represents the heat conducted between the die/workpiece through their interface. T_d and T_w are die and workpiece temperatures, respectively, and h_{lub} is the heat transfer coefficient for the lubricant. The last term is the contribution of the heat generated due to friction along the die-workpiece interface, q_f being the surface heat generation rate due to friction.

6.3.2. Coupled Thermomechanical Analysis Procedure

Generally, for nonisothermal forming processes, a coupled thermo-viscoplastic analysis is carried out wherein it is necessary to simultaneously solve the material flow problem (for a given temperature distribution) and the heat transfer equations. The coupling between the deformation analysis and the heat transfer analysis is given through the the material constitutive relationship for the workpiece. This is because for a rigid-thermoviscoplastic material, the flow stress is a function of effective strain, effective strain rate, temperature and microstructure. Further, the problem of thermal analysis consists of two parts, thermal analysis of the workpiece, and that of the die. The two are connected through the boundary conditions imposed at the die workpiece interface. It is assumed that the die is rigid and not subject to any deformation during the process. This greatly reduces the number of equations to be solved simultaneously, which in turn reduces the cost of the solution. There is also no heat generation in the die and therefore deformation calculations are not necessary.

Many researchers have determined the velocity field, either experimentally or by simulation, and then used the velocity field to determine the internal heat generation. The velocity and temperature fields are not coupled in these analyses. One approach to solving the combined thermomechanical problem is to alternate between the deformation and thermal analyses iteratively until both solutions converge. The deformation analysis gives the nodal velocities (from which strain and strain rate may be calculated), and the thermal solution of the workpiece and die yields their respective temperature distributions. Assuming a converged solution has been obtained for all the above quantities at some time t_i (the subscript referring to the time

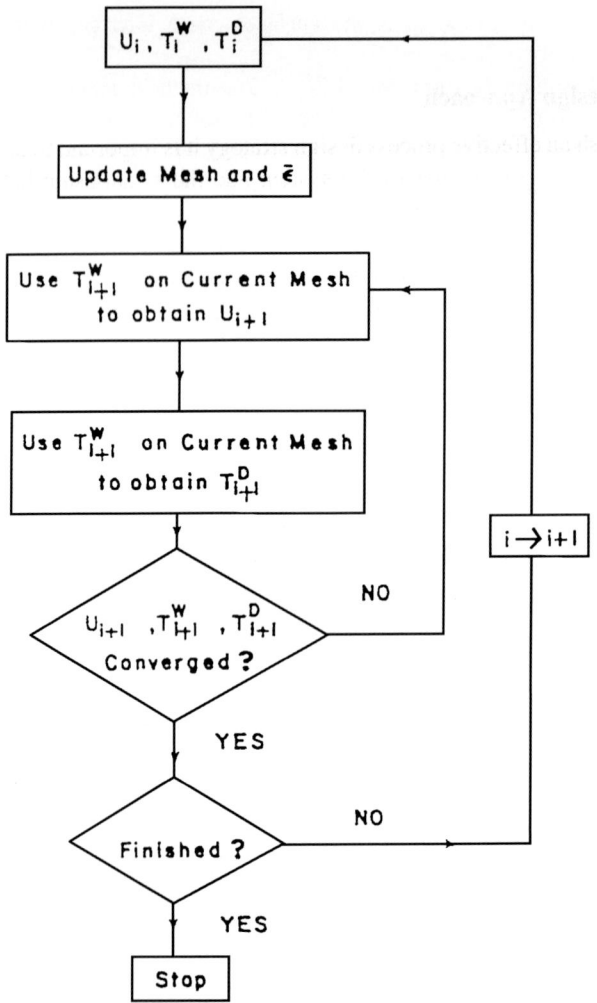

Figure 7.   Coupled thermal and deformation analysis.

step), the next step is to obtain converged solutions for the velocities and temperatures at time step $t_{i+1} = t_i + \Delta t$ after updating the mesh configuration of the workpiece at the end of step t_i. The process is repeated until the required ram stroke has been reached. This is illustrated in the flowchart in Figure 7, where, U, T^W, and T^D refer to the velocity field, workpiece temperature field, and die temperature field, respectively, and i is the current time step number.

6.4. OPTIMIZATION-BASED DESIGN STRATEGY

6.4.1. Design Approach

To establish an effective process design strategy it is important to understand the essential characteristics of the system and the relationship between its various interacting parameters and field variables. The previous sections described in detail some of the process parameters and field variables associated with the forging system, and their influence on the quality of the deformation process in general. This section describes the treatment of the forging system as an optimization problem for designing optimal process parameters. The general strategy and approach for solving the optimization problem using mathematical programming techniques is also discussed.

Optimization of most metal forming processes using the FE method is a complicated procedure because of the incremental formulation used in modeling the complex, unsteady nature of metal flow. In the forging system, generally, the geometry of the boundary changes with stroke. In addition, the stiffness matrix and load vector are functions of the nodal velocities of the system, resulting in a nonlinear set of finite element equations. The analysis path of the process is thus traced in an incremental manner by considering a series of discrete equilibrium states each corresponding to a specific value of time. The geometry configuration is known and is updated at every time step, while the velocity and temperature distributions are determined in an implicit iterative manner, using the Newton-Raphson or direct iteration method, until a converged solution is obtained.[24]

From the finite element (FE) system at every time step an optimization model is built and solved. In this context the FE system includes the physical model, the boundary conditions associated with it, and the interacting process variables. As the process simulation normally requires a number of time steps before completion, to optimize the entire domain of the forging process a number of optimization cycles are required. There is thus continual interaction between the finite element system and the optimization system during the process of optimization at every time step.

The finite element simulation is started with suitable input parameters to the system. The simulation is then 'frozen' at every time step (or predetermined time steps), wherein the optimization system extracts information from the FE system to set up and solve the optimization problem in an iterative manner. The optimal values are then supplied back as inputs to the finite element system and simulation is continued for the next time step of the process. The procedure is repeated, and the simulation and optimization are carried out in sequence until the required stroke is reached. This generic procedure is explained by means of a flow chart in Figure 8.

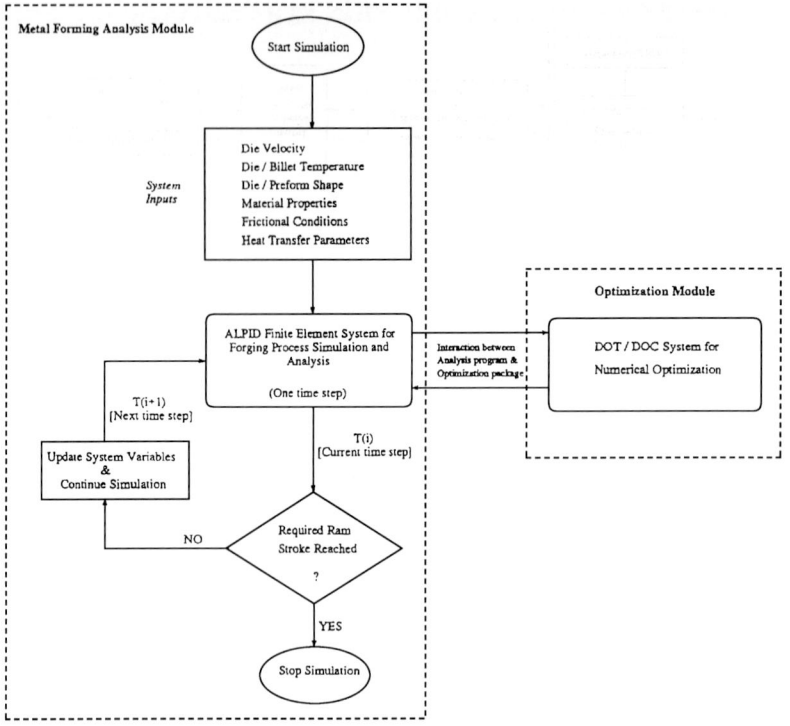

Figure 8.   Generic process design flowchart.

In this work, the nonlinear rigid viscoplastic finite element program ALPID (Analysis of Large Plastic Incremental Deformation)[43] is modified and used for simulation and analysis of the forging process. The program DOT (Design Optimization Tool)[45] is used for implementing the actual optimization procedure. A finite difference approach is used for gradient calculations for sensitivity analysis. The optimization program DOC (Design Optimization Control)[46] is used for defining the optimization problem and for constructing objective/constraint function approximations to reduce the computational cost involved. DOC and DOT are general purpose optimization programs developed by VMA Engineering, Inc., and are generally used as 'black box' programs to be linked with the FE analysis code for performing design optimization and/or sensitivity analysis.[45,46] These programs have a number of built-in algorithms and options for performing optimization, some of which are treated in detail in the following sections.

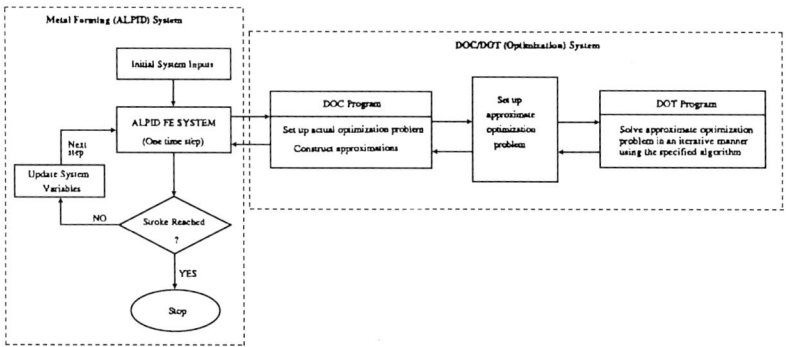

Figure 9. Interaction between the ALPID, DOT, and DOC systems.

Figure 9 shows the nature of interaction between the ALPID, DOT and DOC systems.

6.4.2. Mathematical Optimization Model

Most design processes require the minimization or maximization of some parameter which may be called the design objective. For the design to be acceptable, it must satisfy a variety of limitations which are referred to as constraints. Relative to the design space, the constraints actually define hyper-surfaces that encircle the set of acceptable, or so-called feasible or admissible designs.[47] Optimization problems are nonlinear in general. It is therefore necessary to employ iterative numerical solution schemes and determine the optimum design through a sequence of redesign and reanalysis.

In the most general sense, numerical optimization solves the nonlinear, constrained problem; Find the set of design variables, X_i, $i = 1, N$, contained in vector X, that will

$$\text{Minimize} \quad F(X) \qquad (4.1)$$

Subject to;

$$g_j(X) \leq 0 \quad j = 1, M \qquad (4.2)$$

$$h_k(X) = 0 \quad k = 1, L \qquad (4.3)$$

$$X_i^l \leq X_i \leq X_i^u \quad i = 1, N \qquad (4.4)$$

Equation (4.1) defines the objective function which depends on the values
of the design variables, X. Equations (4.2) and (4.3) are inequality and
equality constraints (which are also functions of the design variables X)
respectively, and Eq. (4.4) defines the region of search for the minimum. The
bounds defined by Eq. (4.4) are referred to as side constraints. N, M, and
L are the number of design variables, number of inequality constraints, and
the number of equality constraints, respectively. The superscripts l and u
refer to the lower and upper bounds on the design variable.

In this work, the ram velocity and initial die temperature are identified
and selected as the essential parameters to be designed, as they have considerable influence on the thermomechanical deformation characteristics of
the workpiece for the entire duration of the process. As explained earlier,
the design is carried out in a step-wise manner while the initial die temperature is a single value to be input (to the system) at the start of the simulation.
This quantity is therefore not changed after the simulation has begun and
is not directly used as a design variable for optimization. Instead, in addition
to the die velocity, a new parameter called the initial die temperature
adjustment parameter is introduced and used as the design variable. This
parameter is treated as a continuous system variable and is designed and
updated (if required) at every simulation time step. A detailed explanation of the role of the design variables in the design process is given in
Section 6.5.

Constraints are imposed on the rate of material deformation within the
workpiece, and on the die-workpiece interface boundary temperatures. To
obtain uniform material deformation and distribution of properties in the
workpiece, the elemental effective strain rate variance and/or nodal temperature variance are chosen as the cost functions depending on the design
requirement.

6.4.3.   Sensitivity Analysis[48]

Solution of any optimization problem involves the determination of a search
direction along which the cost function decreases. This in turn requires the
calculation of the derivatives of the objective and constraint functions with
respect to the design variables. The design sensitivity analysis procedure is
used to calculate the gradient of a function with respect to the design
variables and is an integral part of any optimization process. Sensitivity
analysis is responsible for a large portion of the total computational cost
during optimization. It is therefore important to use an efficient algorithm
for evaluating the sensitivity derivatives. When using finite element programs the gradients are generally calculated as part of the analysis process itself. But, if explicit closed form equations relating the design variables
to the objective and constraint functions are not available, semi-analytical

or finite difference methods are resorted to. These approximate methods, though relatively easy to implement, are computationally very expensive and tedious.

In this work, the central difference approximation is used for obtaining the sensitivity derivatives. This method is more accurate than the first order finite difference methods and results in reduced truncation errors.[49,50] As an example, using the central difference method, the sensitivity derivative (i.e., first derivative) of the strain rate variance, $\dot{\varepsilon}_{var}$ (objective function), with respect to the die velocity (design variable), V_d, is given as follows. Since $\dot{\varepsilon}_{var}$ is a function of the die velocity, V_d, we can write, $\dot{\varepsilon}_{var} = f(V_d)$. Then,

$$\frac{d\dot{\varepsilon}_{var}}{dV_d} = \frac{f(V_d + \Delta V_d) - f(V_d - \Delta V_d)}{2\Delta V_d} = \frac{\dot{\varepsilon}_{var(V_d + \Delta V_d)} - \dot{\varepsilon}_{var(V_d - \Delta V_d)}}{2\Delta V_d} \quad (4.5)$$

where, $\Delta V_d = h$ is called the step size, and is the length of the interval over which the approximation is made. ΔV_d represents a change in the die velocity, and $f(V_d + \Delta V_d) = \dot{\varepsilon}_{var(V_d + \Delta V_d)}$ represents the strain-rate variance value corresponding to this change. As is evident from the above equation, the centered difference approximation actually uses information that is equally spaced (i.e., one step forward and one step backward) around the point where the derivative is estimated. Since no closed form relationship between the strain rate variance and the die velocity is readily available, the finite element program ALPID is used as a 'black box' to determine the sensitivity of $\dot{\varepsilon}_{var}$ to the change in die velocity. A similar procedure is used for determining the other sensitivity derivatives during the course of the optimization process. The finite difference interval size influences the accuracy and rate of convergence of the optimization problem. The program has a built in criterion to determine the interval size, and as such, the finite difference interval may vary from one iteration to another and from one FE time step to the other, to enable greater accuracy and efficiency during implementation.

6.4.4. Use of Approximations in Optimization[48]

Two major difficulties are associated with the process of interfacing a finite element analysis package with an optimization program. The first is a programming difficulty. Optimization codes typically expect subroutines that evaluate the objective function and constraints. When the analysis program is large or the source code of the program is not available, it is very difficult to transform the entire analysis package into a subroutine called by the optimization program. The second problem is the high computational cost associated with using the analysis program as a 'black box'. In many cases evaluation of the objective function and constraints requires the execution of costly finite element analyses for determining the system response quantities. The optimization process may require the evaluation of the objective function and gradients many times, and the cost of repeating the FE analysis so many times is generally prohibitive in nature.

The sequential approximate optimization approach suggested by Schmit and Farshi[51] overcomes both these problems. The computational cost is cut down by the use of approximate analyses during portions of the optimization process. The analysis package is first used to analyze an initial design, and then to generate information that allows the construction of approximation functions representing the objective function and constraints. Explicit or global approximations may be constructed by carrying out analysis at a number of points in the design space and using the response at these points to construct a polynomial approximation to the response at other points. The optimization package may then be applied to the approximate problem. The DOC program uses this approach for building approximations.

But even this may not be a viable option when a large number of design variables are involved. It is more common to use local approximations based on the derivative of the objective function and constraints with respect to the design variables. The simplest approach is to replace the objective function and constraints with linear approximations based on these derivatives. However, these approximations are valid only in the neighborhood of the design space. Therefore, it is necessary to impose limits, called move limits, on the magnitudes of changes in the design that are permitted while the approximate analysis is used. Following an optimization based on approximate analysis and move limits, an exact analysis is performed at the design point obtained by the approximate optimization, and new derivatives are calculated so that a new approximation for the objective function and constraints may be constructed. The process is repeated until convergence is achieved, typically measured by the magnitude of changes in the objective function or the degree of satisfaction of the optimality conditions.

In this work, a linear approximation is used to relate the strain rate dependent quantities (objective and constraints) to the die velocity, while a conservative approximation is used to relate these quantities to the initial die adjustment parameter and die fillet radius correction factor. The temperature dependent quantities are related to all the design variables by means of conservative approximations. These approximations were decided upon after considerable parametric study involving the evaluation of their accuracy and their applicability to metal forming related systems.

In linear approximations the constraint function (for e.g.) is approximated using a Taylor series expansion. For a given function $g(X)$ this may be written as,[48]

$$g_L(X) = g(X_0) + \sum_{i=1}^{n} (x_i - x_{0_i}) \left(\frac{\partial g}{\partial x_i} \right)_{X_0} \qquad (4.6)$$

This approximation is accurate for all functions which behave linearly. But, for nonlinear problems it is inaccurate even for small changes in the design variable. Accuracy can be increased by including additional terms in the Taylor series expansion, but involves costly calculations in evaluating higher order derivatives of the function.

If the function does not behave in a fairly linear manner, a conservative approximation is used. This 'hybrid' approximation has the advantage of both, linear and reciprocal approximations, and is given as,[47]

$$g_c(X) = g(X_0) + \sum_{i=1}^{n} C_i(x_i - x_{0i}) \left(\frac{\partial g}{\partial x_i} \right)_{X_0} \qquad (4.7)$$

where,

$$C_i = \begin{cases} x_{0i} / b_i, & \text{if } x_{0i}(\partial g / \partial b_i) \leq 0 \\ 1 & \text{otherwise} \end{cases} \qquad (4.8)$$

Here, $C_i = 1$ corresponds to a linear approximation and corresponds to a reciprocal approximation in x_i.

6.4.5. Approximate Optimization Problem

The general optimization problem setup is given in Section 6.4.2. Once the approximate functions (Section 6.4.4) have been constructed, the approximate optimization problem is formulated. The move limits specified define the domain in which the approximate optimization problem is solved. The approximate optimization problem is given as,

$$\text{Minimize } f_L^k(X) = f(X_0) + \sum_{i=1}^{n} (x_i - x_{0i}) \left(\frac{\partial f}{\partial x_i} \right)_{X_0} \qquad (4.9)$$

Subject to:

$$g_c(X) = g(X_0) + \sum_{i=1}^{n} C_i(x_i - x_{0i}) \left(\frac{\partial g}{\partial x_i} \right)_{X_0} \leq 0$$

$$(x_i)^{L_k} \text{ or } (x_i)^L \leq x_i \leq (x_i)^{U_k} \text{ or } (x_i)^U, \quad i = 1, 2, \ldots, N \qquad (4.10)$$

where $(b_i)^{L_k}$ is the local lower limit allowed on the design variable and is specified as $(1.0 - ML) * b_i$, ML is the percentage of move limit specified for the design variable at the design point, and $(b_i)^U_k$ is the local upper limit on the design variable and is specified as $(1.0 + ML) * b_i$. The terms $(b_i)^L$ and $(b_i)^U$ are the global lower and upper limits on the design variable. The value of ML is very critical in solving any approximate problem.

6.4.6. Implementation of the Optimization-based Methodology

The flowchart in Figure 10 describes the approach used in setting up and solving the forging process optimization problem. The design procedure followed is explained in the following steps:

Step 1:   Initialize the input (design) variables for performing metal forming analysis and optimization.

Step 2:   Perform the analysis at the initial design using ALPID (for one time step).

Step 3:   Extract the objective and constraint function values from the ALPID analysis. These functions are generally in terms of the field variables like effective strain rate and and nodal temperature, and are explained in detail in Chapter 5.

Step 4:   Determine gradients of the objective and constraint functions with respect to the design variables using a finite difference approach. This requires repeated analyses using ALPID at the current time step.

Step 5:   If the design cycle is greater than 1, check for convergence of the actual optimization problem; if it is converged, stop. Otherwise proceed to step 6. "Design cycle" here refers to the entire process involving extraction of data from ALPID for determining the objective/constraints, building approximations, and solving the approximate optimization problem in an iterative fashion. Convergence is based on satisfaction of the constraints, and change in the objective function in consecutive cycles of the process.

Step 6:   Construct approximations for the cost (objective) and constraint functions.

Step 7:   Solve the approximate optimization problem using the DOT optimization package. The modified feasible directions algorithm is used in obtaining the optimal solution.

Step 8:   Increment the design cycle count.

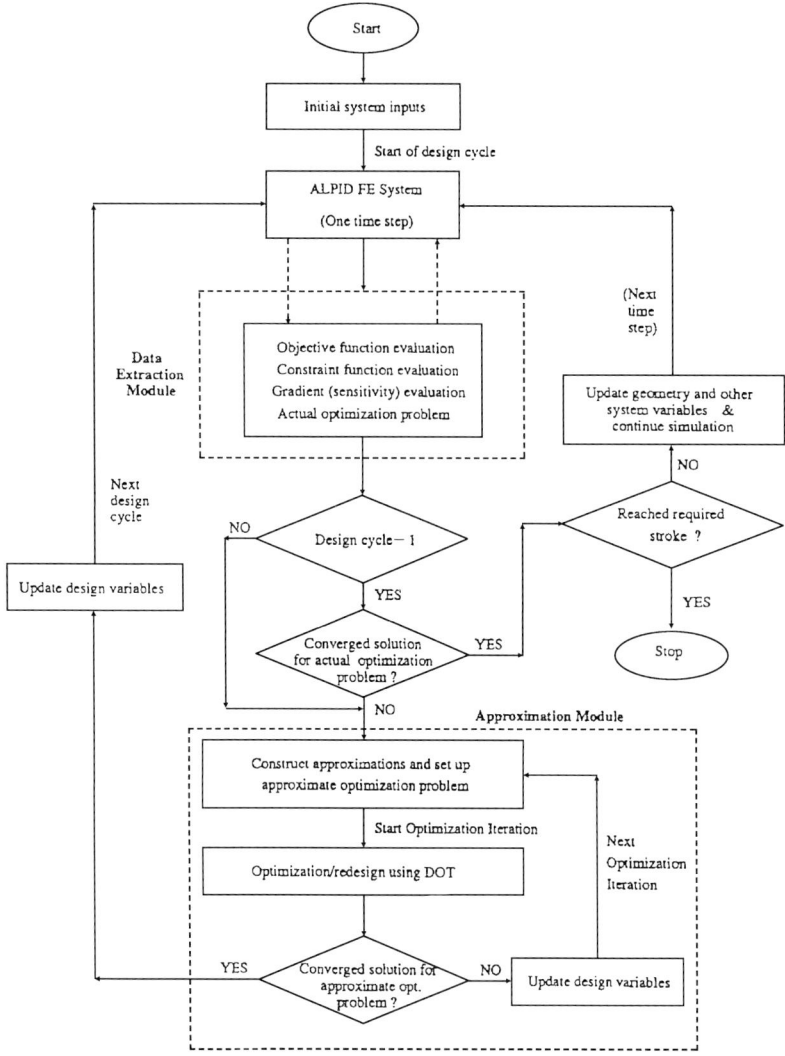

Figure 10. Forging process simulation/optimization flowchart.

Step 9: Go back to step 2 and continue the process.

Step 10: If global convergence (i. e., convergence of the actual optimization problem) is obtained, update the system variables and workpiece geometry, and continue FE simulation for the next time step.

Step 11: Repeat steps 1–10 until the required stroke is reached.

It may be observed that at each finite element simulation time step an independent optimization problem is formulated and solved to obtain optimal process parameters. This could be very time consuming and computationally intensive if the FE model is large and the simulation involved a large number of time steps. To overcome this problem a method has been devised to determine suitable points along the stroke of the ram where optimization should be performed. The method is based on the change in boundary condition of the FE system and is explained in detail at a later stage in this report.

6.4.7. Optimization Algorithm/Method Used

The modified method of feasible directions[45,47] is used in solving the approximate optimization problem. Figure 11 pictorially explains this algorithm which is one of many methods provided in DOT for performing design optimization. The overall optimization process involves the following steps:

Step 1: Start with an initial guess, i.e., $X = X^0$, and set the increment counter to zero, i.e., $q = 0$

Step 2: Start the increment counter $q = q + 1$

Step 3: Evaluate the objective function $F(X^{q-1})$ and the constraint $g(X^{q-1})$.

Step 4: Calculate the gradients of the objective function and the critical constraint $\nabla g(X^{q-1})$.

Step 5: Determine the search direction, S^q.

Step 6: Perform a one-dimensional search to find the value of α^*.

Step 7: Update the design variable set, $X^q = X^{q-1} + \alpha^* S^q$.

Step 8: Check if the problem has converged to the optimum; if converged, then exit from the optimization module. Otherwise, go to step 2 and repeat the procedure.

6.5. FORGING PROCESS PARAMETER DESIGN USING NUMERICAL OPTIMIZATION

It is important to study and control the metal flow behavior during forging to ensure uniformity of deformation and good die-filling. In addition, control of local strain rates in the deforming workpiece helps in the development

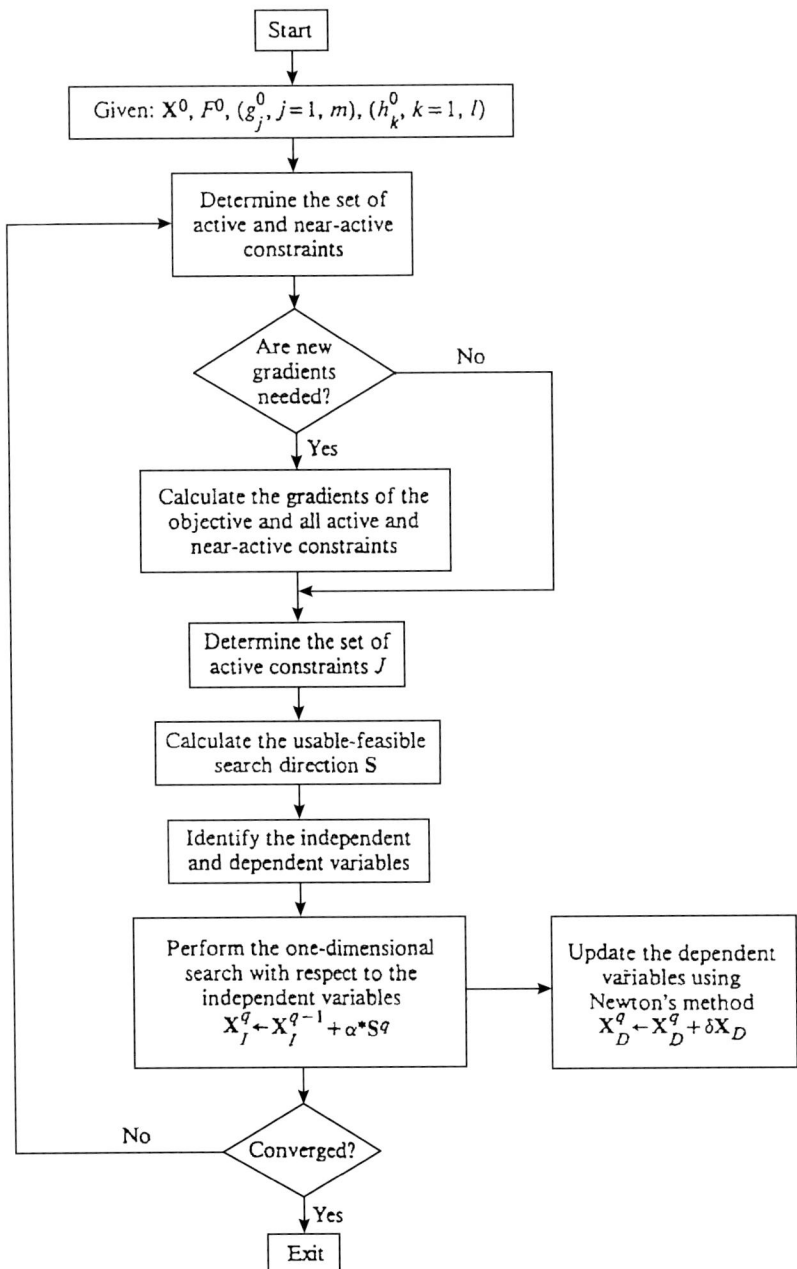

Figure 11. The feasible directions algorithm.

of 'favorable' mechanical and microstructural properties in the finished product.[33,39]

This goal can be achieved by optimizing forging system parameters like die velocity and initial die temperature based on pre-determined deformation and temperature related process requirements. An overview of the optimization based design methodology was given in the previous section. This section explores in more detail the nuances involved in the design of optimal die velocity schedules and initial die temperature using the proposed methodology.

6.5.1. Die Velocity Design for Strain Rate Control

Die velocity is an important parameter associated with the forging process and influences to a large extent the thermo-mechanical characteristics of the deforming workpiece. In this section different aspects of monitoring and controlling elemental effective strain rate by means of the die velocity are explored. Effective strain rate of any element in the finite element model is directly dependent on the nodal velocities associated with it; it is therefore an instantaneous quantity and is directly influenced by changes in the die velocity. However, there are some system dependent constraints on how effectively the strain rate may be controlled while simulating the forging process. The metal working process, though nonlinear in nature, is treated in a piecewise continuous manner by linearizing it at every simulation time step as explained in Sections 6.2 and 6.3. The deformation analysis therefore does not allow a large change in effective strain in one time step of the simulation. This constraint is represented by a maximum strain increment limit per step in the analysis program. In addition, there are other constraints such as the forging press acceleration/deceleration capability, and, the shape of the die. But it has earlier been shown that the strain rate in the billet is strongly dependent on the die velocity[33] and may be controlled at least to some extent by suitably adjusting the ram velocity as the process is carried out.[40-42]

A design methodology based on numerical optimization has been developed to design optimal ram velocity schedules that meet the design requirements on the strain rate. Generally, the design objective is to maintain the effective strain rate at some desired value (or within a range of desired values), while in some cases certain regions of the billet may be required to follow a known strain rate trajectory. It may not always be possible to instantaneously bring the strain rate to the required value due to inherent limitations in the modeling of the system and/or process. The required strain rate is thus gradually approached and maintained (if required) by means of the simulation-optimization cycle.

To generate the optimal die velocity schedule, an optimization problem is formulated with die velocity as the design variable. Constraints are placed on the rate of material deformation, which in this context means the effective strain rate of a particular element or a group of elements in the workpiece model. Uniformity of deformation is always a critical factor in the quality of the final product and in the distribution of properties through the material being deformed. The variance of strain rate in the workpiece domain is thus chosen as the objective function, and it is sought to minimize this quantity by performing numerical optimization. In the most generic sense this may be stated mathematically as,

$$\text{Objective: Minimize} \quad (\dot{\bar{\varepsilon}}_{\text{var}}) \tag{5.1}$$

$$\text{Subject to:} \quad \dot{\bar{\varepsilon}}_{l_j} \leq \dot{\bar{\varepsilon}}_{r_j} \leq \dot{\bar{\varepsilon}}_{u_j}$$

$$\text{Design Variable:} \quad V_d, \quad V_d^l \leq V_d \leq V_d^u \tag{5.2}$$

where, V_d is the die velocity, and, V_d^l and V_d^u are certain lower and upper bounds on the die velocity, determined based on the process requirement and the forge press capabilities. By suitably varying V_d^l and V_d^u at every time step, we may also control the change in velocity at every time step, thus placing a constraint on the acceleration/deceleration of the forge press. Here, $\dot{\bar{\varepsilon}}_{\text{var}}$ is the volume weighted strain rate variance in the billet given by,

$$\dot{\bar{\varepsilon}}_{var} = \frac{\sum\limits_{i=1}^{n} (\dot{\bar{\varepsilon}}_i - \dot{\bar{\varepsilon}}_{avg})^2 V_i}{\sum\limits_{i=1}^{n} V_i} \tag{5.3}$$

$\dot{\bar{\varepsilon}}_{avg}$ is the volume weighted average strain rate in the workpiece, given by,

$$\dot{\bar{\varepsilon}}_{avg} = \frac{\sum\limits_{i=1}^{n} \dot{\bar{\varepsilon}}_i V_i}{\sum\limits_{i=1}^{n} V_i} \tag{5.4}$$

Also, $\dot{\bar{\varepsilon}}_{r_j}$ is the strain rate of the selected critical element(s) in the billet. $\dot{\bar{\varepsilon}}_{u_j}$ and $\dot{\bar{\varepsilon}}_{l_j}$ are the upper and lower limits (respectively) within which the strain rate of the selected element must lie. $\dot{\bar{\varepsilon}}_{r_j}, \dot{\bar{\varepsilon}}_{l_j}$, and $\dot{\bar{\varepsilon}}_{u_j}$ are generally determined from available experimental data or from material processing maps. The subscript 'j' represents the number of elements whose strain rate paths are to be controlled, and hence is a measure of the total number of constraints in the optimization problem. As described above, each value of 'j' results in two constraints. V_i is the volume of element 'i' in the workpiece. In all the above cases the summation index is 'n', where, 'n' represents the number of elements in the workpiece finite element model.

The die velocity, V_d, influences the objective and constraint functions considerably and is thereby chosen as the design variable, as mentioned earlier. Closed form analytical relationships between the die velocity (design variable) and strain rate variance/elemental strain rate (objective/constraint functions) are not available. The relationship between these quantities is implicit in nature as far as the finite element system is concerned. So, for the present, the finite element program, ALPID, is used as a 'black box' for determining the influence of the design variable(s) on the objective/constraint functions.

Solution of this optimization problem, in effect, results in an optimal die velocity schedule which forces the effective strain rate (of the selected element) in the billet to satisfy the strain rate constraint(s). In other words, by optimizing the die velocity at every simulation time step, in addition to minimizing the strain rate variance, the strain rate in the billet may be maintained at any given value or constrained between two specified values, based on the way the constraints are posed. For e.g., an equality constraint may be used for maintaining the effective strain rate in the deforming workpiece at a constant value. If the constraint on strain rate is dynamic in nature, i.e., it varies at every simulation time step, then, the same methodology may be used to track a given strain rate trajectory rather than just maintain the strain rate at some given value.

The optimization process generally requires a trial vector of design variables to be input before the iterative solution procedure is started. For nonlinear problems the solution is dependent to some extent on the initial guess value. In this work, since the FE analysis at every time step is treated and solved as an independent optimization problem, an optimized die velocity is obtained at the end of each simulation time step. This velocity is then supplied as the initial guess for the next time step optimization process, rather than using some arbitrary trial value. This is done to maintain continuity in the die velocity profile for the entire simulation. Also, using the optimized ram velocity from the previous step as initial guess for the current step (optimization process) would probably lead to faster convergence of the optimization process. This can be explained in the following manner. If small simulation time steps are used, the thermomechanical behavior of the deforming workpiece would not change drastically from one time step to the next unless there is a change in boundary condition of the system. As such, the optimized velocity from the previous time step would be close to the optimal ram velocity for the current time step, resulting in fewer optimization iterations (before convergence) for the current time step.

A key aspect in the implementation of this procedure is the selection of the so-called 'control' element(s). The 'control' element is generally chosen based on its location in the workpiece relative to the die, its state of deformation relative to the stroke, and the number of different deformation modes it is likely to go through before the simulation is completed. In numerical terms this would mean that elements more sensitive to changes in the process parameters and elements undergoing large deformations (in any given time step) are good candidates for selection as 'control' elements. In many instances it may be required to maintain the maximum strain rate in the billet below a certain specified value. In such cases, since the element having maximum strain rate in the workpiece domain is likely to vary with progress in deformation, the 'control' element also changes with stroke. The methodology developed has a built-in strategy to transfer "control" from one element to another as and when required. Also, elements on the boundary are influenced to a large extent by the accuracy in modeling of the die as well as the frictional and heat transfer conditions prevalent at the die-workpiece interface. Many forming defects are also initiated at the surface of the workpiece. The behavior of these elements is therefore unpredictable to some extent and not truly representative of the state of deformation over the entire billet domain. As a result, the boundary elements are not generally chosen as 'control' elements for the optimization process. Also, in some cases it may be required to monitor and control the strain rate of a group of elements rather than just one element. Using the above methodology this would be a fairly simple task and involves the addition of extra strain rate constraints associated with the additional 'control' elements. During multi-element control it is very important that the constraint values be chosen very carefully. Otherwise, the optimization problem may not converge because of one or more constraints being violated at the end of the optimization iterations.

Implementation of this methodology is likely to improve the performance of the forging process to a large extent. A reduction in the strain rate variance in the workpiece implies a more uniform flow of material, which in turn reduces the possibility of the occurrence of defects. Controlled strain rate trajectories would also result in obtaining desired (favorable) microstructures and service properties in the final product. Earlier research in this area used advanced control schemes to achieve the same objectives.[40–42] but the optimization approach because of its simplicity and ease of implementation is probably more appealing.

6.5.2. Die Velocity Design for Temperature Control

Temperature plays an important role in metal forming processes and strongly influences the microstructural properties of the deforming workpiece. Temperature control is a much more difficult and complicated issue compared to strain rate control. From Section 6.3 it is observed that nodal temperatures are influenced by a number of heat generation and heat transfer terms. The first term in Eq. (3.9) (deformation heat generation) is dependent on the type of forging process used. This term plays a more important role in high speed forging processes like mechanical press forging (about 20 in/sec ram velocity in general) or hammer (about 80 in/sec ram velocity) forging. For these processes changing the die velocity during the course of the process is not viable because of the high inertia of the die and the very short processing time involved. Therefore, in this study we confine ourselves to forging processes utilizing hydraulic presses having relatively lower ram velocities (for instance below 5 in/sec). Earlier, Boer et al.[33] designed a varying die velocity profile for processing NIM80A material and implemented it on a hydraulic machine with a microcomputer as the controller.

During forging, generally, temperatures within the workpiece are functions of the heat generated due to plastic deformation, which in turn depends on the rate of material deformation and the die velocity. Temperatures on the boundary surface are more influenced by the initial die/billet temperatures and the resulting die chilling effect. The last four terms in Eq. (3.9) mainly affect the boundary surface temperature. Among these terms, the radiation heat term is neglected because it has an insignificant magnitude when compared to the other terms. Again, if the process does not last very long, the convection heat loss is not very significant. At the boundaries, the friction heat has some influence on the temperature, but is still not the dominant term. The dominant factor is the heat conducted between the billet and die through their interface. Because the die temperature is usually lower than that of the billet, the heat transfer due to conduction between these two bodies is considerable, and a severe temperature gradient exists at the die-workpiece boundary. This phenomenon is called the die chilling effect. Severe thermal stresses induced at the die-workpiece interface could lead to 'thermal softening' of the die and adversely affect die life. The die chilling effect also affects the microstructure and mechanical properties of the workpiece as it causes a non-uniform distribution of temperature in the billet. Therefore, in this study, one of the primary goals was to develop a method to alleviate the die chilling effect. Since this phenomenon occurs mainly at the boundaries, only the control of boundary nodal temperatures is taken up in this work. The difficulties involved in boundary nodal temperature control stems from the fact that it is influenced by the initial die/billet temperature also in addition to the die velocity.

In this work, the objective during temperature control is to minimize the nodal temperature variance in the workpiece domain. Alternatively, the temperature range in the workpiece may also be minimized. These two objectives serve the same purpose and are desirable because they would result in a more uniform distribution of mechanical and service properties in the final product. Since temperature is not easily controllable, for the scope of this work the goal is confined to keeping the nodal temperatures above a specified value rather than trying to maintain them at any given value. To achieve the above objectives, the optimization scheme described earlier is again resorted to. A constraint is placed on the lowest boundary (die-contacting) nodal temperature, henceforth called the 'critical' node. The logic behind this is if the lowest boundary temperature is maintained above a certain value, all the other boundary nodal temperatures would also be above the specified value. If this nodal temperature is to trace a specified temperature path, then dynamic constraints are imposed. These constraints vary from one time step to the next as the simulation progresses, thereby forcing the temperature of the 'critical' node to conform to the specified trajectory. The dynamic constraints are determined from a known polynomial approximation of the required temperature as a function of the ram stroke. The key issue here is in the determination of the polynomial approximation which is explained in the section on initial die temperature design.

In mathematical terms the optimization model associated with the temperature control problem may be defined as,

$$\text{Objective:} \quad \text{Minimize} \quad T_{var} \quad \text{or} \quad (T_{max} - T_{min}) \tag{5.5}$$

$$\text{Subject to:} \quad T_{l_j} \le T_{r_j} \le T_{u_j} \tag{5.6}$$
$$\text{Design Variable:} \quad V_d, \quad V_d^l \le V_d \le V_d^u$$

where, V_d is the die velocity, and, V_d^l and V_d^u are certain lower and upper bounds on the die velocity, determined based on the process requirement and the forge press acceleration/deceleration capabilities. By suitably varying V_d^l and V_d^u at every time step, we may also control the change in velocity at every time step, thus placing a constraint on the velocity change of the forge press. T_{var} is the variance of nodal temperature in the entire billet domain, and is given by,

$$T_{var} = \frac{\displaystyle\sum_{i=1}^{m}(T_i - T_{avg})^2}{(m-1)} \tag{5.7}$$

and, T_{avg} is the average strain rate in the workpiece, given by,

$$T_{avg} = \frac{\sum_{i=1}^{m} T_i}{m} \qquad (5.8)$$

$T_{\max}$ and $T_{\min}$ refer to the maximum and minimum values of nodal temperatures in the billet. The difference between these two quantities gives the temperature range in the workpiece domain. T_{r_j} is the temperature of the selected 'critical' node(s) in the billet. T_{u_j} and T_{l_j} are the upper and lower limits (respectively) within which the temperature of the selected node must lie. These values are again determined from material processing maps or from available experimental data. The subscript 'j' represents the number of nodes whose temperature paths are to be controlled, and hence is a measure of the total number of constraints in the optimization problem. In all the above cases the summation index 'm' represents the total number of nodes in the workpiece finite element model.

Since die velocity is used as the design variable in the above design problem the elemental strain rates (which are more sensitive to changes in die velocity than nodal temperatures) may acquire undesirable values. To account for this, the above optimization problem may also be defined with additional strain rate constraints, and with the dual objective of minimizing, both, the temperature and strain rate variances in the workpiece, as discussed later in this report.

During nonisothermal forging process design, the strain rate control (constraint) and the temperature control (constraint) may have a conflicting requirement on the die velocity. For example, the strain rate control may need the die velocity to be reduced; at the same time the temperature control may actually require the die velocity to be increased. In such a situation the temperature constraint is laxed a little and redefined. The strain rate control is given higher priority because it is more sensitive to changes in the die velocity. Therefore, in this design scheme, if the above mentioned conflict occurs, the die velocity is decided based on the strain rate. Actually, temperature is an accumulated quantity and is therefore not easily controllable at every time step. The main contribution of the optimal die velocity design to the thermal aspect of the process is in reducing the temperature range (and/or variance) in the deforming billet by reducing the die-workpiece contact time. This may not always be possible, though, and a more effective way to directly influence the interface temperature is by designing appropriate initial die/billet temperatures for the process.

6.5.3. Initial Die Temperature Design

Because of the strong die chilling effect during hydraulic press forging, the most efficient way to influence the boundary nodal temperatures and reduce the temperature range in the workpiece is by designing an appropriate initial die temperature. But, the optimization based design procedure described earlier is a step-wise process while initial die temperature is a single value and needs to be specified at the beginning of the process. To overcome this problem, the following design approach is proposed.

In this scheme, a new design variable called the die temperature adjustment parameter (ΔT_d) is introduced. By adding ΔT_d to the present die temperature (T_d) the total die temperature is given by ($T_d + \Delta T_d$), and the conduction term in Eq. (3.9) may now be written as,

$$\int_{S_l} h_{lub} (T_d + \Delta T_d - T_w) \mathbf{N}_T dS \qquad (5.9)$$

By using this equation instead of Eq. (3.9) in solving the heat transfer equations the influence of the die temperature adjustment parameter on the nodal temperatures may be effected and evaluated. Here, it is assumed that the change in die temperature does not have a significant effect on the heat transfer terms due to convection and radiation.

Adding ΔT_d to the die temperature actually changes the temperature gradient at each step by a certain amount. The other terms in Eq. (5.9) are mainly shape functions and heat transfer coefficients and do not change much during the whole process. For a case in which the temperature gradient needs to be reduced by the same amount at each time step (by applying the ΔT_d correction), nearly the same ΔT_d value may be expected from the optimization scheme. This suggests that to achieve the design objective, the actual initial die temperature should be changed by the average value of ΔT_d for all the time steps. The optimization problem for initial die temperature design may be stated as,

$$\text{Objective:} \quad \text{Minimize} \quad T_{var} \quad \text{or} \quad (T_{max} - T_{min}) \qquad (5.10)$$

$$\text{Subject to:} \quad T_{l_j} \leq T_{r_j} \leq T_{u_j}$$
$$\text{Design Variable:} \quad \Delta T_d, \quad \Delta T_d^l \leq \Delta T_d \leq \Delta T_d^u \qquad (5.11)$$

Based on the simulation-optimization scheme explained earlier, the die temperature adjustment parameter (ΔT_d) is designed at every time step of the simulation. But, unlike in the case of die velocity design, the die temperature adjustment parameter is not used to update the actual die temperature at every simulation step. This would affect the thermal aspects of the present simulation and would result in an inaccurate temperature design for the remaining simulation time steps. This stems from the fact that temperature, unlike strain rate (in die velocity design), is a history based (accumulated) quantity. To offset the effect of adding the die temperature adjustment parameter to the die temperature during sensitivity analysis, at the end of optimization for any given time step, ΔT_d is set to zero and the step is repeated before going onto the next simulation step. So, the optimal ΔT_d value at every time step is only stored and not actually passed on to the next simulation step. An arbitrary ΔT_d value is then used as initial guess for optimization in the next time step.

Solving this optimization problem results in one optimal ΔT_d value for every simulation step. The average of all these values is taken as the 'effective' die temperature adjustment. This quantity is then added to the actual (initial) die temperature to obtain the optimal initial die temperature value.

$$\Delta T_{d(avg)} = \frac{\sum_{k=1}^{p} \Delta T_{d(k)}}{p} \tag{5.12}$$

where k is the simulation step count and p is the total number of time steps for the simulation to be completed. The optimal die temperature for the process is then given by,

$$T_{d(optional)} = T_{d(initial)} + \Delta T_{d(avg)} \tag{5.13}$$

where, $T_{d(optional)}$ is the optimal initial die temperature and $T_{d(initial)}$ is the initial die temperature with which the simulation was started. The simulation is then repeated with the optimal initial die temperature value as explained in Section 6.6.2.4.

Here again if the 'critical' node is required to track a specified temperature trajectory, dynamic constraints have to be imposed, as explained earlier. The key issue now is how to generate a desired nodal temperature profile which not only satisfies the design requirement, but also gives nearly the same temperature gradient change at each design step. Through parametric studies it was found that for hydraulic press forging, the temperature curve of a die-contacting node is nearly a straight line. Therefore, we may with reasonable accuracy use a 3rd order polynomial to approximate the temperature path of a particular node with respect to time. Then, by suitably changing the coefficient of the first order term in the polynomial, the desired temperature profile may be generated. The above design scheme works successfully if the die temperature is not very close to the billet temperature, and if the die velocity does not vary by a significant amount. These conditions hold good in most hot die forging cases. The detailed procedure for designing the initial die temperature is explained in the numerical examples section of this report.

6.5.4. Integrated Forging Process Parameter Design

The earlier sections described the design of die velocity and intial die temperature on an individual basis. Each of these parameters influences the forging process in a different manner. To obtain truly optimal values for these parameters with respect to both deformation and temperature aspects of the forging process, and to study their combined effect on the process variables, the integrated and simultaneous design of these process parameters needs to be carried out. This can be very easily done using the optimization-based methodology developed, by combining all of the optimization problems described earlier into one thermomechanical optimization problem given in mathematical terms as,

$$\text{Objective:} \quad \text{Minimize} \quad W_1(\dot{\bar{\varepsilon}}_{var}) + W_2(T_{var}) \quad (5.14)$$

Subject to:

$$\dot{\bar{\varepsilon}}_{l_j} \leq \dot{\bar{\varepsilon}}_{r_j} \leq \dot{\bar{\varepsilon}}_{u_j} \quad (5.15)$$

$$T_{l_j} \leq T_{r_j} \leq T_{u_j} \quad (5.16)$$

Design Variables: $\quad V_d,\ \Delta T_d$

Side constraints:

$$V_d^l \leq V_d \leq V_d^u$$

$$\Delta T_d^l \leq \Delta T_d \leq \Delta T_d^u$$

All the above terms have been explained earlier and are not elaborated upon in this section. As explained in Section 6.5.2, the temperature range could be substituted in place of the temperature variance as the second objective function. The weighting factors for the objective function (W_1 and W_2) may each vary between 0 and 1 such that their total equals 1. These factors need to be carefully chosen based on the design requirement as they affect the convergence of the optimization problem significantly. Multiobjective optimization problems often have trouble converging because of the large difference in the magnitudes of the objective functions being weighted. This is also true in the above problem where one of the objectives is in terms of the elemental strain rate and the other is described in terms of the nodal temperature. Often this problem is handled by restructuring the optimization problem and retaining only the most important objective, while placing the rest of the weighted objectives as constraints. In the present case, $\bar{\dot{\varepsilon}}_{var}$ is retained as the objective while suitable constraints are placed on T_{var} which is appended to the constraint set instead.

Solution of the above optimization problem results in an optimal die velocity schedule and optimal initial die temperature for the process, thereby enabling the control of strain rates and nodal temperatures in the workpiece domain. In addition, minimization of the strain rate variance and temperature variance in the workpiece improves the uniformity of deformation and results in uniform distribution of mechanical and microstructural properties in the deforming material with minimal defect formation.

6.6. NUMERICAL EXAMPLES

A general purpose computer program has been developed to implement the design techniques described in this work. This program is built on the rigid viscoplastic finite element program ALPID (Analysis of Large Plastic Incremental Deformation). In addition to using ALPID subroutines for analysis purposes, an optimization based design and control module has been developed and included in the program. The modified ALPID program has been interfaced with optimization programs DOC (Design Optimization Control) and DOT (Design Optimization Tool) to perform mathematical optimization for designing optimal forging process parameters based on specified deformation and temperature related constraints. The program developed has the following capabilities:

1. ALPID simulation for both isothermal and nonisothermal processes.
2. Optimal die velocity design for isothermal process.
3. Optimal die velocity design for nonisothermal process.
4. Initial die temperature design for nonisothermal process.
6. Any combination of the above designs.

This chapter presents a few numerical examples to support and substantiate the methodology presented in earlier sections of the report. Based on the way the optimization problem is posed, different design requirements and objectives may be met.

6.6.1. Example 1: Upset Forging

The optimization-based methodology developed in this work is demonstrated for an open-die isothermal forging process, wherein, the uniaxial compression of a cylindrical billet is considered. A non-strain hardening material having the constitutive relation $\sigma = 10\dot{\bar{\varepsilon}}^{0.1}$ is used to model the flow stress of the material. The frictional condition at the die-workpiece interface is modeled using a constant shear friction factor of 0.3.

Forging simulation is carried out using two flat dies. It is assumed that the top and bottom die are moving with the same velocity, but in opposite directions. Taking advantage of the symmetry of the problem, only a quarter section of the billet is modeled and analyzed. The finite element package I-DEAS (Integrated Design Engineering Analysis Software) is used for mesh generation during modeling. The initial billet is modeled using 50 isoparametric quadrilateral elements. Deformation of the workpiece is carried out starting with a die velocity of 1.0 inch/sec, for a total ram stroke of 0.6 inch. The entire simulation comprised 100 time steps each of step size 0.01 sec. Figure 12 shows the finite element model of the workpiece before and after deformation.

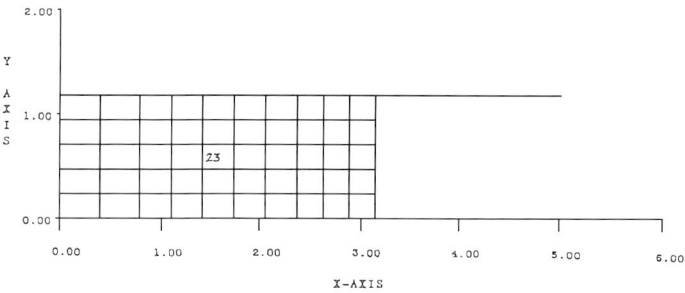

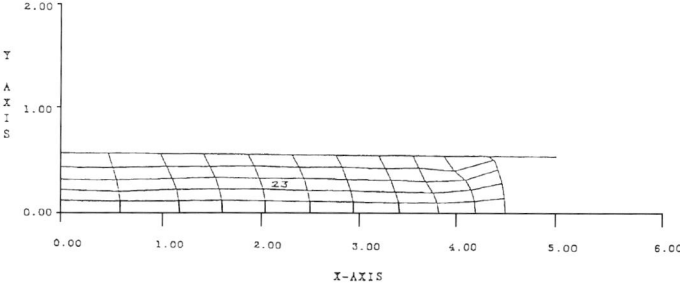

Figure 12. Simple upset forging — FEM simulation.

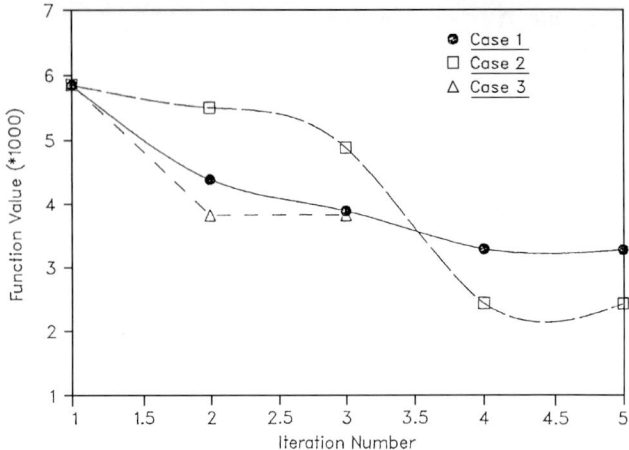

Figure 13.   Objective function iteration history.

At each simulation time step an optimization problem is constructed and solved. Approximations were not used during optimization in this example. To demonstrate the working of the optimization algorithm, the iteration history for one time step of the simulation is presented in Figure 13 for three different cases, using different constraints.

The objective of the optimization process is to minimize the strain rate variance in the billet by designing optimal ram velocity profiles. Constraints are imposed to maintain the strain rate of any given element (or group of elements) at a specified value or within a specified range of values. Element 23 is chosen as the 'control' element. A systematic criterion for the selection of critical elements needs to be developed. In this case the selection of the 'control' element is only for demonstration of the methodology. The change in objective function with every optimization iteration (in one time step of the simulation) is used to evaluate the effectiveness of the optimization algorithm. In case 1 (Figure 13), the strain rate of element 23 (here, $\dot{\bar{\varepsilon}}_r = \dot{\bar{\varepsilon}}_{23}$) is constrained between $\dot{\bar{\varepsilon}}_u = 0.75/s$ and $\dot{\bar{\varepsilon}}_l = 0.65/s$. The corresponding optimization problem is posed as:

$$\text{Minimize} \quad (\dot{\bar{\varepsilon}}_{var})$$

$$\text{Subject to:}$$

$$0.65 \le (\dot{\bar{\varepsilon}}_{23})$$

$$(\dot{\bar{\varepsilon}}_{23}) \le 0.75$$

$$\text{Design variable:} \quad V_d$$

Figure 13 plots the objective function value against the number of optimization iterations before convergence.[45,46] It may be observed that for case 1 the objective function $(\bar{\dot{\varepsilon}}_{var})$ decreases gradually, and convergence is obtained in 5 iterations. In case 2, the constraint limits are changed to $\bar{\dot{\varepsilon}}_u = 0.84\,/\mathrm{s}$ and $\bar{\dot{\varepsilon}}_l = 0.56\,/\mathrm{s}$, while all other parameters and conditions are kept the same. This in effect increases the feasible solution space and is less demanding on the optimization algorithm. It may be observed (Figure 13) that, again, 5 iterations are required before convergence is obtained. In case 3, an equality constraint is used such that $\bar{\dot{\varepsilon}}_{23} = 0.7\,/\mathrm{s}$. Generally, using such a constraint would narrow down the feasible region and would make optimization more difficult. It is interesting to note, however, that for this case the converged solution is obtained in 3 iterations of the optimization algorithm. This probably suggests that the FE simulation time step is very small, and the objective and constraint functions may be behaving in a linear manner during that time interval. It may be interesting to note that for case 1, 5 function calls and 5 gradient calls are required before convergence is obtained, and a total of 11 finite element analysis calls are required for the entire optimization process.

6.6.1.1. Maintaining the strain rate at a specified value

This section presents the results of optimization for the entire simulation of the uniaxial compression process. The objective of the optimization process, again, is to minimize the strain rate variance within the deforming material. But, the actual goal is to design an optimal ram velocity schedule that would result in strain rates being maintained at specified values.

The FE model, workpiece material, and operating conditions described earlier are again used for this example. The goal is to maintain the strain rate of element 23 in the billet at 0.7/s. The optimization problem may be stated as:

$$\text{Minimize} \quad (\bar{\dot{\varepsilon}}_{var})$$

$$\text{Subject to:}$$

$$0.7 \le (\dot{\varepsilon}_{23})$$

$$(\dot{\varepsilon}_{23}) \le 0.7$$

$$\text{Design variable:} \quad V_d$$

Optimization is performed at every time step of the simulation and the resulting velocity (optimal) is used as input to restart simulation for the next time step. The process is continued until the required stroke of 0.6 inch is

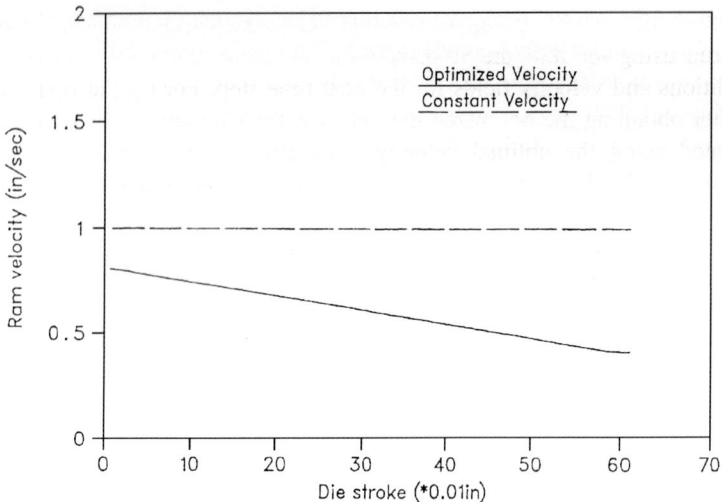

Figure 14. Upset forging — Optimal die velocity schedule.

reached. On an average about 4–5 iterations, involving 10–12 finite element calls, are required at every time step of the simulation.

Figure 14 shows the optimal (designed) ram velocity schedule for this example, while Figure 15 shows the strain rate profile of element 23 corresponding to this velocity. It needs to be mentioned here that for any given

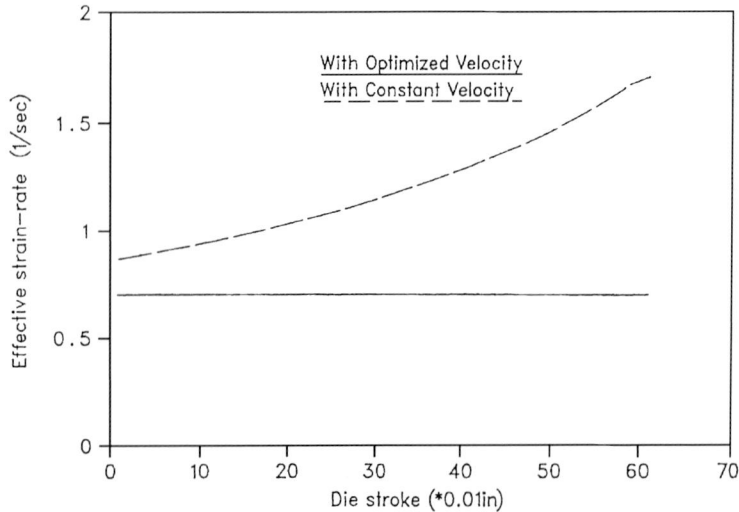

Figure 15. Upset forging — Controlled strain rate trajectory.

time step, after obtaining the optimal die velocity, the current step simulation is rerun using the new die velocity, so as to obtain the correct boundary conditions and velocity fields for the next time step. For e.g., at time step 1, after obtaining the optimized die velocity, the simulation (for step 1) is repeated using the optimal velocity. This also explains why there is a discrepancy in the starting values of die velocity and strain rate for the optimized and and unoptimized cases in Figures 14 and 15, respectively.

From Figures 14 and 15, it may be observed that using a constant die velocity results in the strain rate of element 23 ($\bar{\varepsilon}_{23}$) increasing with progress in deformation. But, using the optimized die velocity profile results in being maintained at the required value. In addition, minimization of the strain rate variance in the billet considerably improves the uniformity of deformation and reduces the possibility of occurrence of defects.

6.6.1.2.   *Tracking a given strain rate trajectory*

In this example the goal is to track an arbitrary strain rate trajectory. Stated differently, the aim is to make the strain rate path of element 23 conform to a specified trajectory. An arbitrary strain rate trajectory (Figure 17) is chosen as the desired path to be tracked in this example. The simulation (FE) model described earlier is again used for this example. The optimization problem is also setup as described earlier. This example differs from the previous one in the way the constraints are posed. In this example, a dynamic strain rate constraint is used, i.e., the strain rate constraint is varied at every

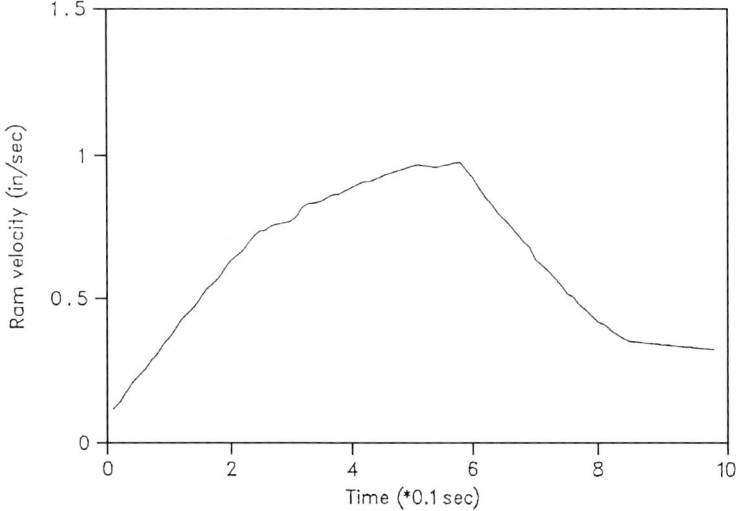

Figure 16. Velocity schedule for the strain rate tracking problem.

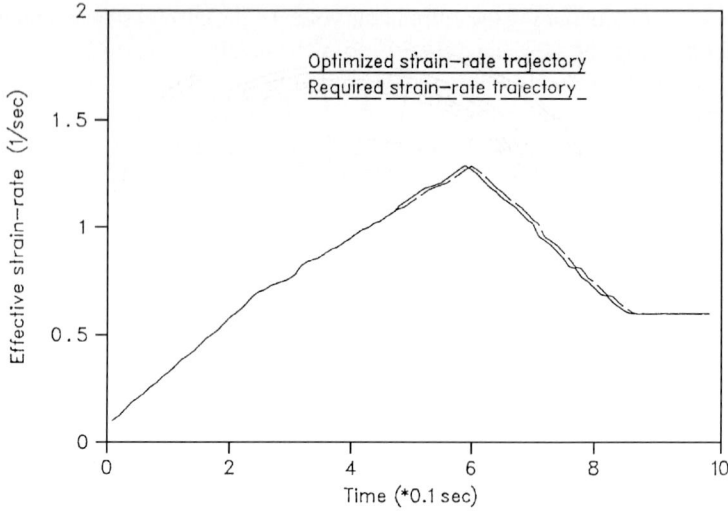

Figure 17. Tracking of strain rate trajectory of element 23.

time step of the simulation, and for each of the corresponding optimization problems. Actually, for this particular example, a series of equality constraints were imposed at every time step of the simulation to ensure that the specified strain rate trajectory is tracked.

Figure 16 shows the velocity profile obtained as a result of using the optimization-based methodology developed, while Figure 17 describes the corresponding $\dot{\varepsilon}_{23}$ trajectory. It may be observed from Figure 17 that the actual strain rate conforms closely with the required strain rate for almost the entire duration of the simulation process.

The methodology developed is thus quite effective in satisfying the strain rate requirements during deformation. In addition, solution of the optimization problem also minimizes the strain rate variance in the workpiece at every time step, resulting in more uniform material flow, and more uniform distribution of properties in the workpiece.

6.6.2.   Example 2: Engine Disk Forging

In this example, optimal process parameters are designed for an axisymmetric engine disk forging (Figure 18) under different operating conditions. The material used for the simulation is Ti-6242 (Titanium alloy), which is a strain rate sensitive material. Due to its symmetry, only a quarter section of the cylindrical billet is modeled. Here again it is assumed that the two dies are

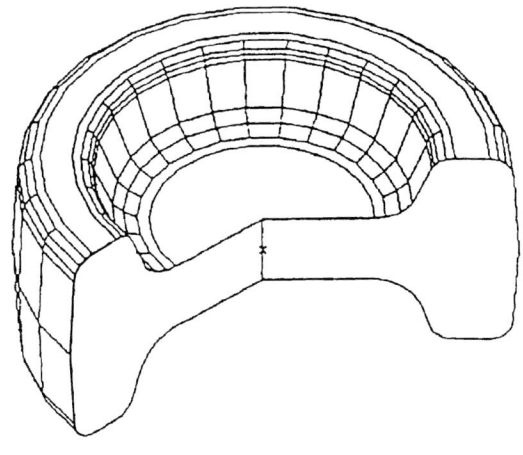

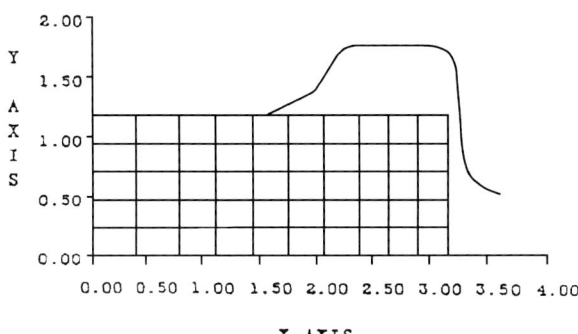

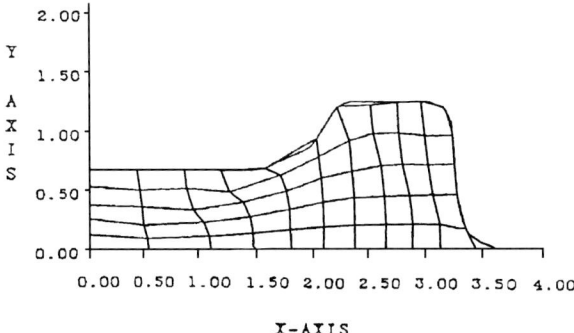

Figure 18. Engine disk forging — FEM simulation.

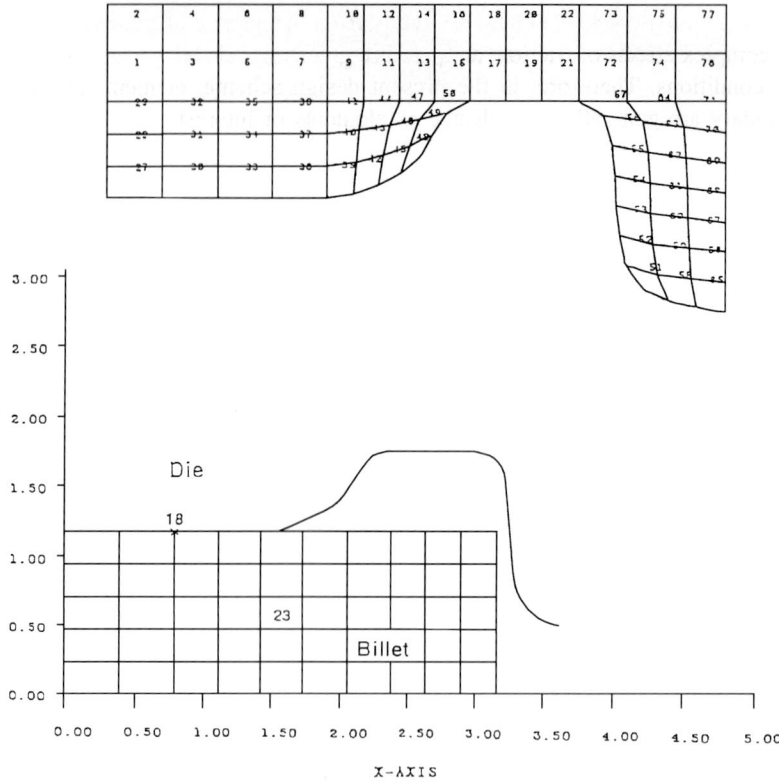

Figure 19. FEM discretization of the billet.

moving in opposite directions with the same die velocity. Discretization of the billet continuum is performed using 50 isoparametric quadrilateral elements and 66 nodes (Figure 19). The die is modeled using 73 finite elements and 100 nodes. A constant shear friction factor of 0.3 is used to model the frictional condition at the die-billet interface. A billet temperature of $1735°F$ and a die temperature of $600°F$ is specified at the beginning of the process. A total die stroke of 0.5 inch is required to complete the forging operation as shown in Figure 18.

Element 23 is selected as the element of interest. The rationale behind this is, due to its location in the workpiece, element 23 is always under deformation and is also likely to go through a number of modes of deformation. As a result, this element is very sensitive to changes in the die velocity and is a good candidate to be the 'control' element. It was observed, however, that boundary elements behave differently from the interior elements due to the complex effects of friction, temperature gradient, and changes in boundary conditions. Therefore, in the present design scheme, elements at the boundary are generally not selected as elements of interest.

6.6.2.1. Isothermal design

This example uses the same engine disk forging model described earlier. An isothermal design is presented which requires the effective strain rate of element 23 to be maintained between 0.45/s and 0.55/s. The starting die velocity is 1.0 in/sec, corresponding to a $\dot{\varepsilon}_{23}$ value of about 0.5/s. The optimization problem is posed as:

$$\text{Minimize} \qquad (\dot{\varepsilon}_{var})$$

$$\text{Subject to:}$$

$$0.45 \leq (\dot{\varepsilon}_{23})$$

$$(\dot{\varepsilon}_{23}) \leq 0.55$$

$$\text{Design variable:} \qquad V_d$$

Figure 20 depicts the optimal die velocity schedule designed for this

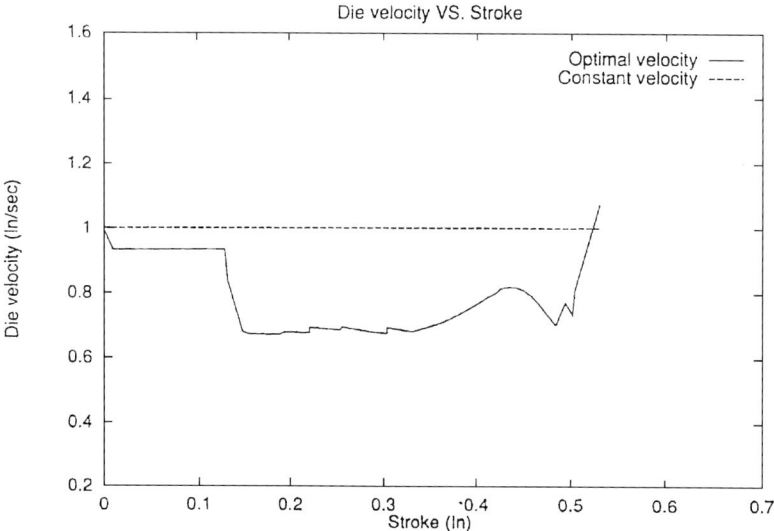

Figure 20.   Isothermal design — Die velocity schedule.

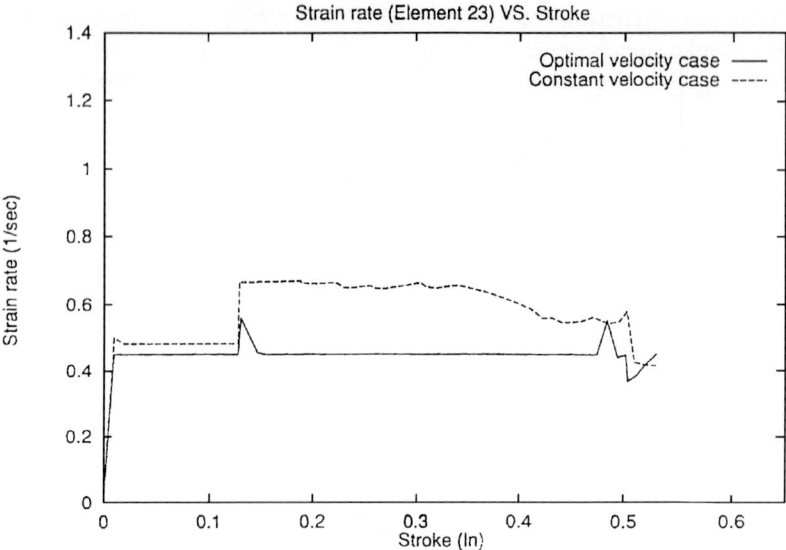

Figure 21. Isothermal design — Strain rate path.

process, and Figure 21 shows the corresponding $\dot{\bar{\varepsilon}}_{23}$ trajectory. During the initial stages of forging the change in $\dot{\bar{\varepsilon}}_{23}$ is relatively small, but around a stroke of 0.13 inch a large jump (nearly a 50% increase) in $\dot{\bar{\varepsilon}}_{23}$ may be noticed. This is because of the sudden change in the deformation mode of the workpiece as it comes in contact with the far end of the die. However, as shown in Figure 21, the design scheme is robust and effective, and immediately reduces the die velocity and brings the effective strain rate back to about 0.45/s in a short time without much oscillation. Near the end of the process though, some instability is observed, causing some fluctuation in both the die velocity and strain rate. This may be attributed to the complex nature of forces that come into play during the die filling process.

In addition to the strain rate of element 23 being maintained at the required value, the strain rate variance is minimized at every time step of the simulation. A sample optimization iteration history showing the variation of the objective function with the iteration number is shown in Figure 22. Figure 23 depicts the strain rate variance history for the entire process. It may be noticed that the variance is considerably lower for the optimal velocity case, resulting in more uniform deformation characteristics.

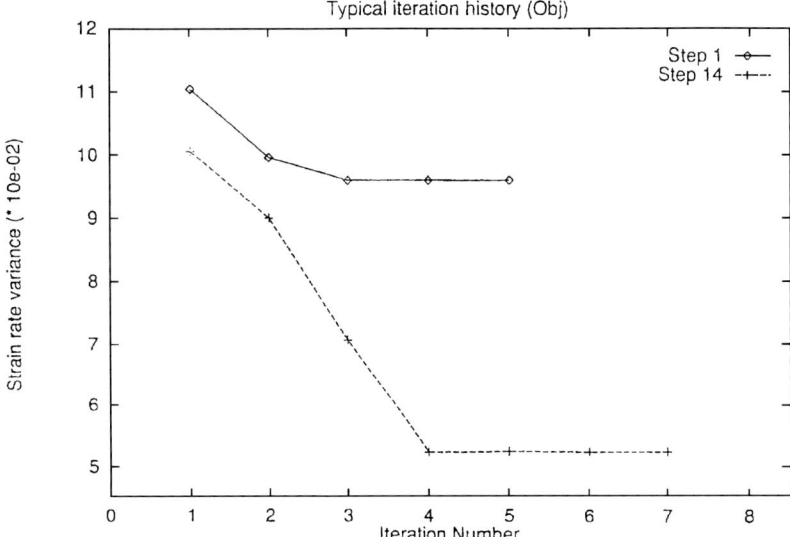

Figure 22.   Isothermal design — Iteration history.

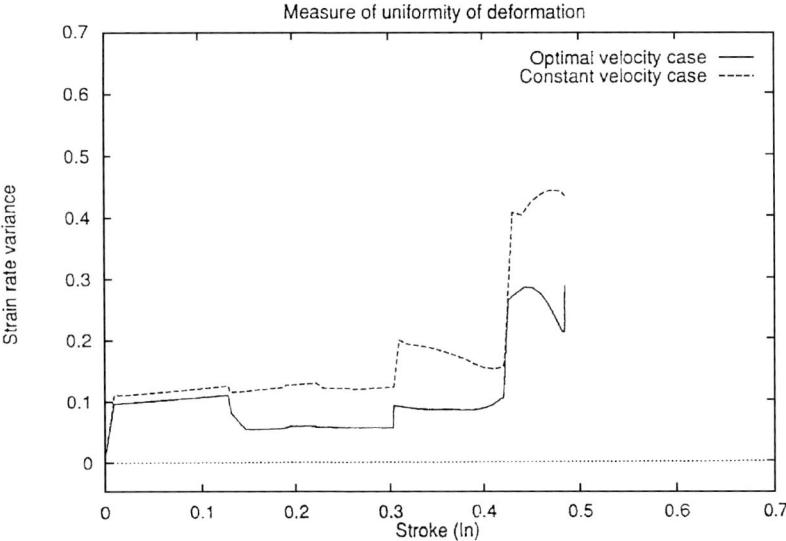

Figure 23.   Isothermal design — Strain rate variance.

6.6.2.2. Temperature gradient reduction

One of the primary objectives in this work is to reduce the temperature gradient at the die-workpiece interface, and thereby, alleviate the thermal stresses induced due to the die chilling effect. This example illustrates the influence of temperature gradient reduction on the boundary nodal temperatures. The design requirements are chosen arbitrarily just to illustrate the method. It is assumed that the temperature gradient at node 18 needs to be reduced, and the effective strain rate of element 23 ($\dot{\bar{\varepsilon}}_{23}$) is to be maintained at 2.0/s. The only design parameter considered is the die velocity. Forging simulation is started with a relatively low die velocity of 0.5 in/sec, which results in $\dot{\bar{\varepsilon}}_{23}$ around 0.25/s. This gives enough room to change the die velocity by large amounts to illustrate the effectiveness of the current method.

Two design cases are studied and presented in this example. The first design, called 'Design I', does not include temperature gradient reduction, and merely involves controlling the strain rate of element 23. The second design, 'Design II', is similar to Design I, but in addition involves a temperature gradient reduction (at node 18) of 3% at every time step. This requirement on the temperature of node 18 is posed as a tracking problem using dynamic constraints as explained below. The optimization objective is to minimize the strain rate variance in the deforming workpiece. The optimization problem may be posed as:

$$\text{Minimize} \quad (\dot{\bar{\varepsilon}}_{var})$$

$$\text{Subject to:}$$

$$2.0 \leq (\dot{\bar{\varepsilon}}_{23})$$

$$(\dot{\bar{\varepsilon}}_{23}) \leq 2.0$$

$$T_{18} \leq T_{des}$$

$$T_{des} \leq T_{18}$$

$$\text{Design variable:} \quad V_d$$

where, T_{des} imposes a dynamic (changes at each time step) constraint on the temperature of node 18. At step i,

$$T_{des} = T_{i-1} + g(1-n)\Delta t$$

where, T_{i-1} is the temperature of node 18 at the previous step $(i-1)$, g is the temperature gradient at step i, n is the constant gradient reduction factor

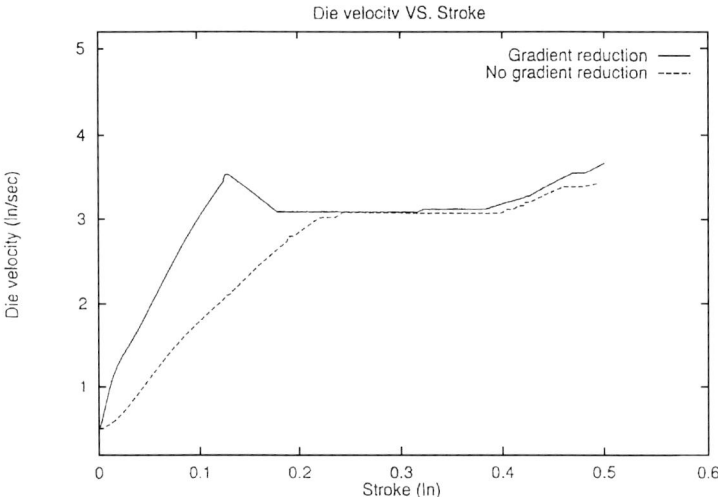

Figure 24.   Temperature gradient reduction — Die velocity profile.

applied at every time step, and Δt refers to the step size. It may be observed that gradient reduction is actually achieved by forcing the temperature of node 18 to follow the path of T_{des}.

The results are presented in Figures 24 through 27. Figure 24 shows the designed die velocity schedule, while Figure 25 presents the controlled

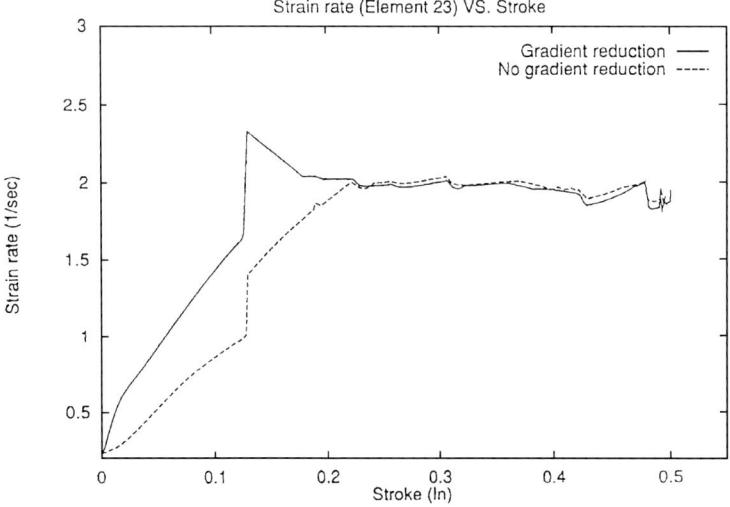

Figure 25.   Temperature gradient reduction — Strain rate trajectory.

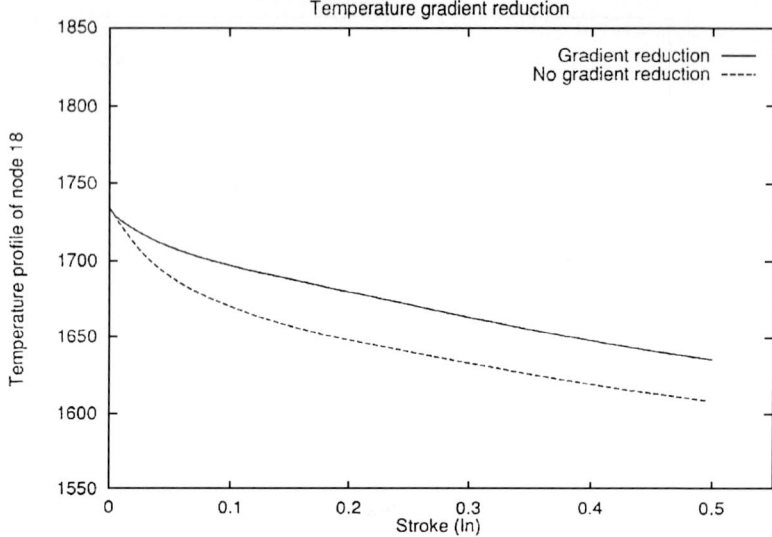

Figure 26.   Temperature gradient reduction — Temperature profile of node 18.

strain rate path of element 23. During Design I die velocity increases at a steady rate till $\dot{\bar{\varepsilon}}_{23}$ reaches 2.0/s at a stroke of 0.23 inch. In Design II, the strain rate increases faster than in Design I because of the effect of temperature gradient reduction. To reduce the temperature gradient the heat generation rate needs to be faster than normal. This can be achieved only by increasing the die velocity. As a result, the strain rate also increases at a faster rate and reaches the required value quicker than in the previous case. When the stroke approaches 0.13 inch, a sudden jump in the strain rate occurs because the workpiece comes in contact with the far end of the die. Consequently, $\dot{\bar{\varepsilon}}_{23}$ overshoots the desired value of 2.0/s. The die velocity is now reduced to bring $\dot{\bar{\varepsilon}}_{23}$ back to 2.0/s. The desired value is finally achieved at a stroke of 0.17 inch. Examining the temperature profiles in Figure 26, it is observed that before a stroke of 0.2 inch, the temperature gradient shows a definite reduction in the second case. But after that, the gradients of both curves are nearly the same because the die velocities in both cases are almost the same. At the end of the process, the temperature of node 18 is $1609°F$ for Design I and $1635°F$ for Design II, a $26°F$ increase in the latter case. It may be concluded that temperature gradient reduction results in an increase in temperature of the die-contacting boundary nodes. It should be noted, though, that this increase is largely a result of the reduced die-workpiece contact time because the die velocity is higher in Design II (for most of the process) than in Design I. In addition, the strain rate variance

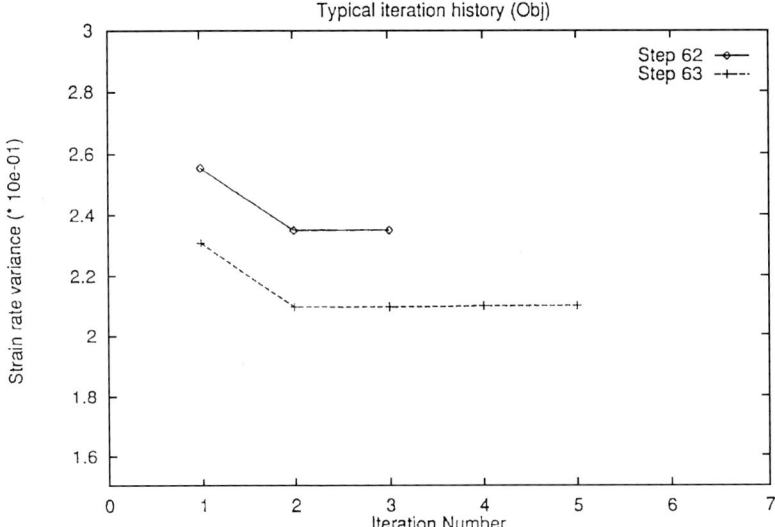

Figure 27.   Temperature gradient reduction — Iteration history.

is minimized resulting in uniform material deformation for the entire simulation. Figure 27 depicts the change in objective function ($\dot{\varepsilon}_{var}$) at different
time steps to demonstrate the effectiveness of the optimization algorithm in
converging to the optimal point.

   This example shows that using the optimization-based design scheme a
suitable die velocity profile may be generated that maintains the strain rate
of critical elements at the desired value and at the same time reduces the
temperature gradient at the die-workpiece interface and the strain rate variance in the workpiece domain.

6.6.2.3.   *Temperature range reduction*

In most hot die forging processes one of the design objectives is to reduce
the temperature range in the billet as much as possible. This improves the
temperature distribution and results in a more uniform distribution of properties in the final workpiece. The following numerical example is presented
to illustrate how the above objective can be met using the proposed design
methodology. The optimization problem may be stated as:

$$\text{Minimize:} \quad (T_{\max} - T_{\min})$$

$$\text{Subject to:} \quad \dot{\varepsilon}_{\max} \leq 2.5 \, / \, s$$

$$\text{Design variable:} \quad V_d$$

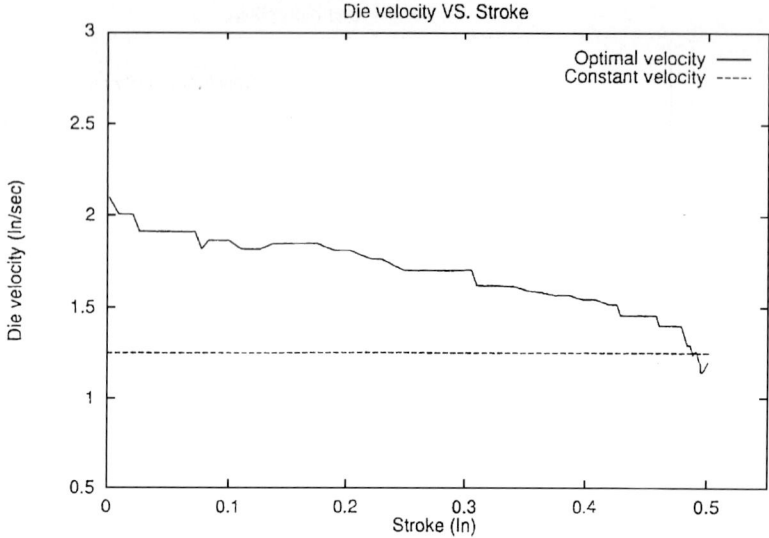

Figure 28.   Temperature range reduction — Die velocity schedule.

Through finite element simulations it was observed that node 18 pos-
sessed the lowest temperature in the billet throughout the process. This is
because node 18 remains in contact with the die from beginning to end, and
the strain rate in the vicinity of node 18 is not very high. The temperature
range (objective function) may now be approximated as $(T_{max} - T_{18})$, where
T_{18} is the temperature of node 18. The strain rate constraint requires the
maximum effective strain rate in the billet to be no more than 2.5/s. The
optimization scheme is used to generate an optimum ram velocity profile.
The optimal results are then compared with the constant velocity results to
demonstrate the effectiveness of the proposed design methodology.

The design scheme has the ability to automatically select the element
possessing the largest strain rate as the element of interest and design an
optimal die velocity schedule based on the corresponding strain rate constraint imposed. Figure 28 gives the ram velocity schedule for the process, and Figure 29 presents the strain rate of the element(s) having the
largest strain rate in the workpiece at any given time step. Starting with a
value of 2.1 in/sec, the designed ram velocity gradually reduces keeping
$\dot{\varepsilon}_{max}$ around 2.5/s for the entire process. The designed ram velocity profile
gives the largest value of die velocity that can be used without violating the
strain rate constraint. The designed profile also gives the minimum contact
time between the die and billet. By minimizing the contact time, the total
heat loss and the temperature range in the workpiece are also reduced. At

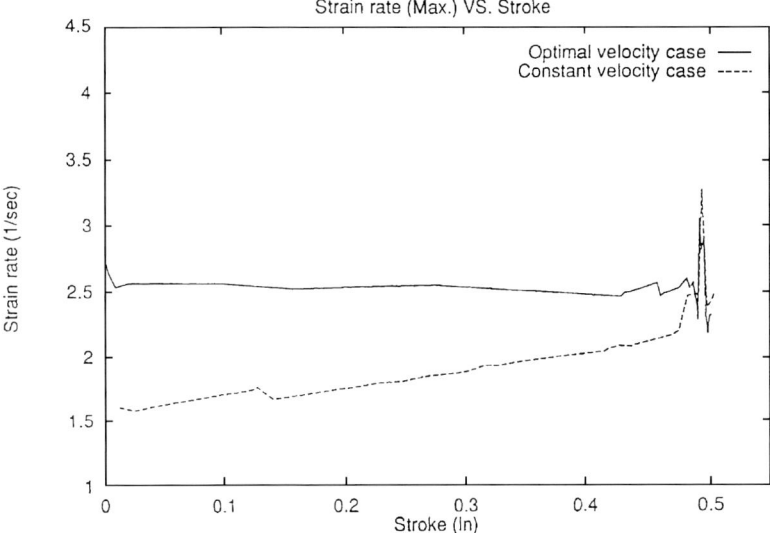

Figure 29. Temperature range reduction — Controlled strain rate path.

the end of the process, node 18's temperature (the lowest temperature in the billet) is $1574°F$ and the maximum temperature in the billet is $1777°F$, giving a temperature range of $203°F$.

For comparison, the above example was also simulated using a constant die velocity. Because $\dot{\varepsilon}_{max}$ gradually increases under a constant die velocity, the largest constant die velocity which can be used without violating the constraint is 1.25 in/sec. By examining Figures 28 and 29 it may be observed that even though for most of the process $\dot{\varepsilon}_{max}$ is far below 2.5/s limit, we can use a constant die velocity no larger than 1.25 in/sec; otherwise $\dot{\varepsilon}_{max}$ would violate the strain rate constraint towards the end of the stroke.

In this example the constant velocity is lower than the optimal ram velocity for most of the forging process (Figure 28). This results in a longer die-workpiece contact time in the former case. At the end of the simulation it may be observed that node 18 has a temperature of $1533°F$, and the maximum temperature in the billet is $1775°F$. The temperature range thus turns out to be $242°F$. Figure 30 clearly illustrates the difference in temperature profiles of node 18 for the two cases. In addition to the above mentioned benefits, for this example, the optimal velocity schedule reduced the process time by about 25.4%. This example shows that by using the optimization-based design scheme, an optimal die velocity profile can be generated to minimize the die-workpiece contact time and the temperature range in the deforming workpiece.

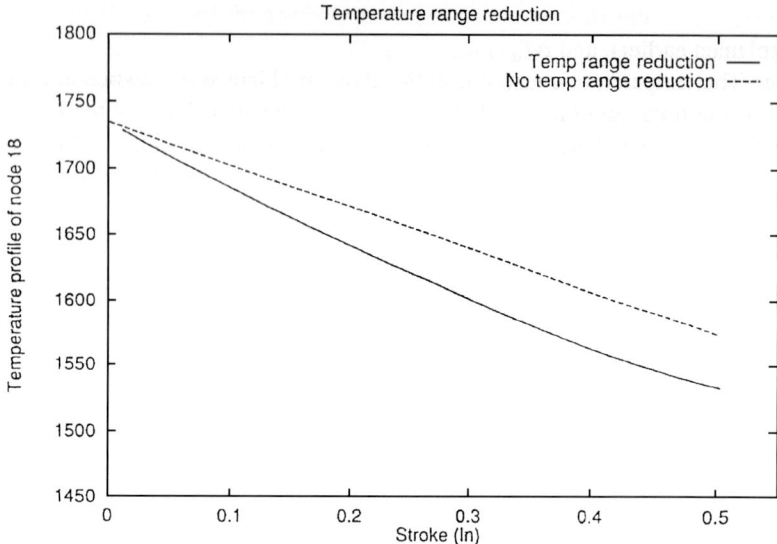

Figure 30. Temperature range reduction — Node 18 temperature profile.

6.6.2.4. *Initial die temperature design*

In forging processes, selection of initial die temperature is an important but difficult decision to make. This numerical example describes how the proposed design methodology could be used to obtain an optimal initial die temperature value. The overall goal is to keep the workpiece (nodal) temperature above a specified value for the entire duration of the simulation. This, in effect, reduces the temperature gradient in the deforming workpiece.

The design requirement in this example is to maintain the billet temperature above $1560°F$ for the entire simulation. It was again observed that node 18 had the lowest temperature in the workpiece for the entire duration of the simulation. Therefore, to keep the billet temperature above $1560°F$ it would be sufficient to ensure that T_{18} is over $1560°F$ for the entire simulation. The design also requires the strain rate of element 23 to be maintained at 0.7/s. The die velocity and initial die temperature adjustment parameter are chosen as design variables. The optimization problem for this example is stated as:

$$\text{Minimize} \quad \dot{\varepsilon}_{var}$$
$$\text{Subject to:} \quad \dot{\varepsilon}_{23} \leq 0.7$$
$$0.7 \leq \dot{\varepsilon}_{23}$$
$$1560 \leq T_{18}$$
$$\text{Design variables:} \quad \Delta T_d \quad \text{and} \quad V_d$$

where $\dot{\varepsilon}_{23}$ is the effective strain rate of element 23 (critical element as explained earlier), and ΔT_d is the initial die temperature adjustment parameter. The temperature constraint in the above problem was satisfied almost all of the time, and the solution seemed to be dictated more by the strain rate requirement. This necessitated a "tracking" approach to the initial die temperature design problem, as explained below. The design is carried out in three steps, each necessitating an independent finite element simulation.

Step 1:   Generating the desired nodal temperature profile

In this step, the design is conducted using only die velocity as the design variable. The initial die temperature used is *600°F*. The corresponding optimization problem is given by,

$$\text{Minimize} \quad \dot{\varepsilon}_{var}$$

$$\text{Subject to:} \quad \dot{\varepsilon}_{23} \leq 0.7$$

$$0.7 \leq \dot{\varepsilon}_{23}$$

$$\text{Design variables:} \quad V_d$$

The purpose of this step is to: (1) Check whether the design requirement can be satisfied without changing the initial die temperature, and (2) If the

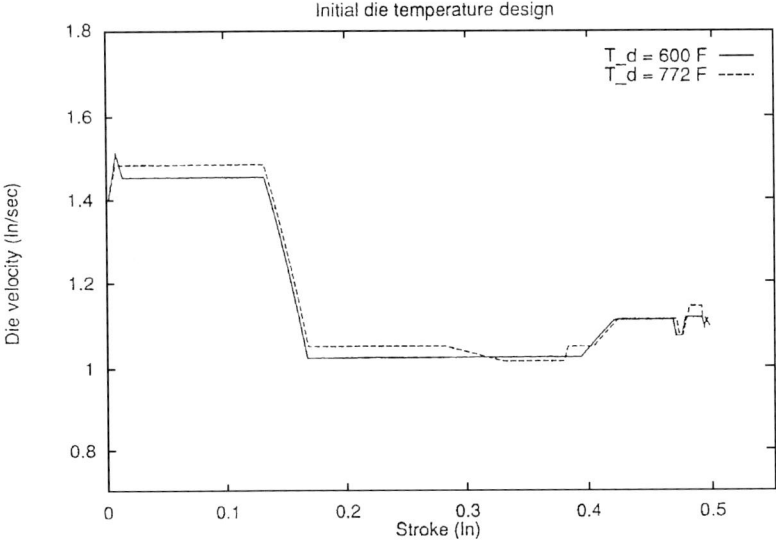

Figure 31.   Initial die temperature design — Die velocity schedule.

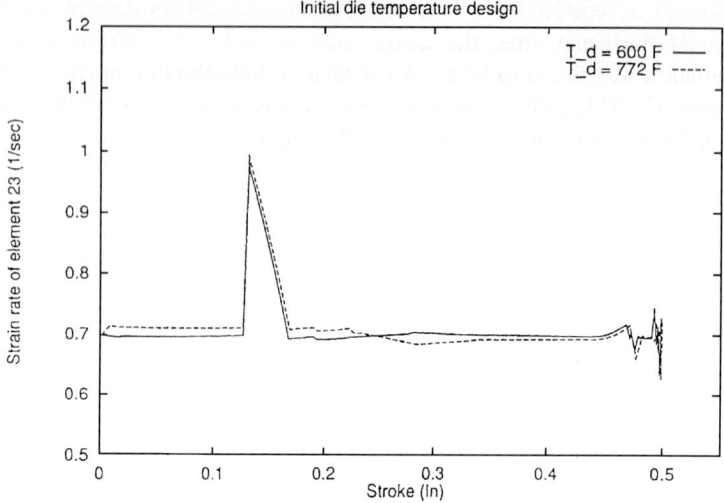

Figure 31.   Initial die temperature design — Controlled strain rates.

design requirement is not met, generate a suitable polynomial representing the temperature path of node 18 that satisfies the specified requirement. The design is started with die velocity, V_d, equal to 1.4 in/sec, corresponding to $\dot{\bar{\varepsilon}}_{23}$ equal to 0.698/s. Figures 30 through 32 give the profiles of V_d, $\dot{\bar{\varepsilon}}_{23}$, and T_{18}, for this example. It may be observed from Figure 31 that the strain rate

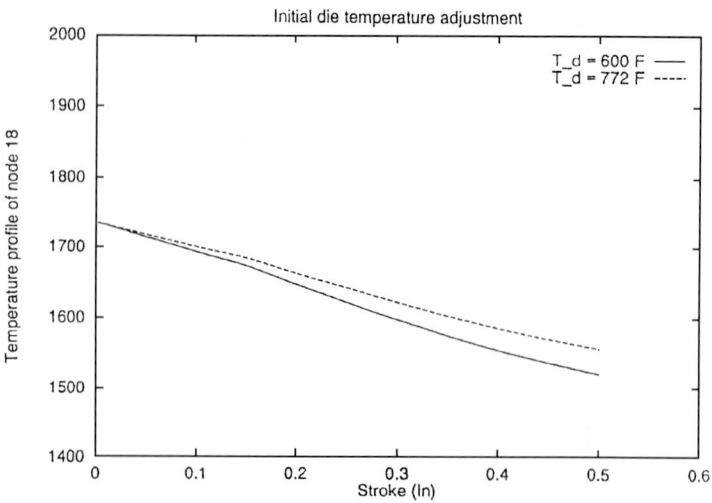

Figure 32.   Initial die temperature design — Node 18 temperature profile.

constraint is satisfied for the entire process under the optimal die velocity (V_d). At the same time, the temperature of node 18 at the end of the simulation was found to be $1519°F$ which violates the design requirement (Figure 32). This calls for an increase in the initial die temperature so as to bring the temperature of node 18 to the required value. To represent the present temperature profile, a 3rd order polynomial is used to fit the path of T_{18} as,

$$T_{18} = 1735 - 641.37t + 355.5t^2 - 25.1t^3 \qquad (6.1)$$

In Eq. (6.1) the first order term is the dominant term influencing the temperature gradient. To raise the nodal temperature, therefore, the coefficient of the first order term has to be changed to an appropriate value. By setting $T_{18} = 1560°F$ (design requirement) at end of the process, the new first order coefficient (from Eq. (6.1)) is calculated as -550. Substituting this back into Eq. (6.1) we have,

$$T_{req} = 1735 - 550t + 355.5t^2 - 25.1t^3 \qquad (6.2)$$

Eq. (6.2) gives the desired temperature profile for node 18 during the initial die temperature adjustment design.

Step 2: Adjusting the initial die temperature

With Eq. (6.2) as the desired temperature profile (for node 18), the design in step 1 is repeated using an additional design variable ΔT_d. The other conditions remain the same as in Step 1. The new optimization problem is stated as follows:

$$\text{Minimize} \quad \dot{\varepsilon}_{var}$$
$$\text{Subject to:} \quad \dot{\varepsilon}_{23} \leq 0.7$$
$$0.7 \leq \dot{\varepsilon}_{23}$$
$$T_{18} \leq T_{req}$$
$$T_{req} \leq T_{18}$$
$$\text{Design variables:} \quad \Delta T_d \quad \text{and} \quad V_d$$

Figures 30 through 32 again give the profiles of V_d and for this example. The value of ΔT_d at every time step is presented in Figure 33. Averaging the ΔT_d values for all the simulation time steps, a mean initial die temperature adjustment value of $172°F$ is obtained (represented by the dashed line

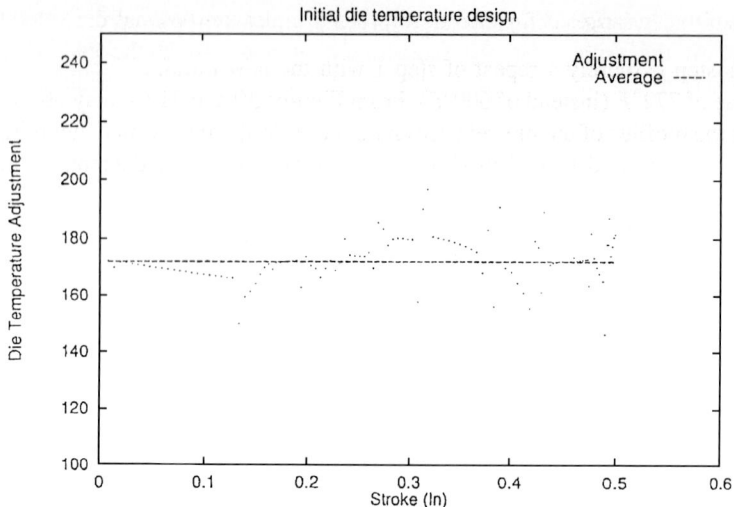

Figure 33. Initial die temperature adjustment parameter.

in Figure 33). The actual initial die temperature which satisfies all the design constraints is thus $772°F$. At each time step, in addition to satisfaction of the constraints, the strain rate variance is minimized, resulting in uniform deformation for the entire process. Figure 34 plots the change in objective function (strain rate variance) against the optimization iteration number at different steps during the simulation.

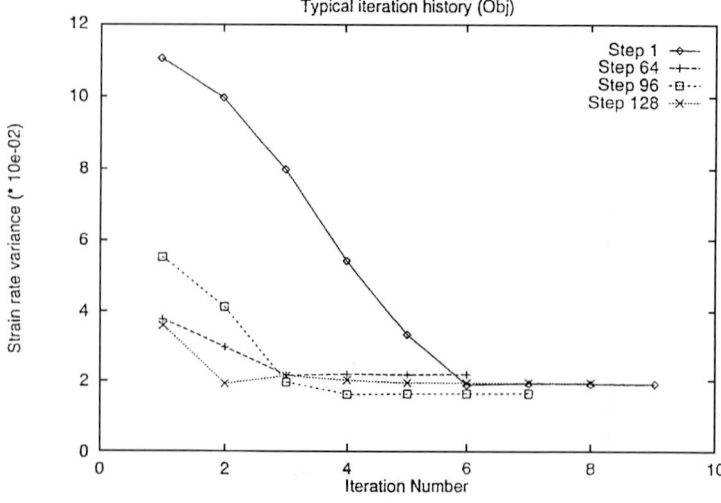

Figure 34. Initial die temperature design — Iteration history.

Step 3: *Redesign using the new initial die temperature*

This step is merely a repeat of step 1 with the new initial die temperature value of $772°F$ (instead of $600°F$). From Figures 30 and 31 we may observe that the profiles of the die velocity and $\dot{\bar{\varepsilon}}_{23}$ are similar before and after initial die temperature design. This shows that changing the initial die temperature does not affect the strain rate of the control element significantly. The temperature of node 18, though, is very sensitive to the change in initial die temperature as observed in Figure 32. With a $172°F$ higher initial die temperature, the final temperature of node 18 is $1557°F$, very close to the design requirement of $1560°F$ (a difference of about 0.2%). The temperature gradient of the boundary nodes (node 18 in particular) is thus reduced. At the same time, it is also observed that the temperature range in the workpiece is reduced from $256°F$ to $218°F$, a drop of about 14%.

   This example shows that the proposed method for initial die temperature (adjustment) design works well, and can be effectively used to reduce the temperature range/temperature gradient in the deforming workpiece. Noting that the increase in boundary nodal temperatures is not linearly proportional to the increase in initial die temperature, solving a design problem such as the one posed in this example would be very difficult and cumbersome without a systematic methodology and/or strategy like the one described in this section. One of the major advantages of the method described here is that the design can be performed in three simulation steps instead of going through a number of trial and error steps as is done normally. This could save a considerable amount of time, effort, and money during the entire design process.

6.6.3. Example 3: IBR Disk Forging

This section presents the design of optimal die velocity schedule for an integrated blade and rotor (IBR) disk forging under nonisothermal conditions.

   The forging of the IBR disk is conducted in two stages, and Figure 35 shows the corresponding billet and die shapes. Figures 35a and 35b show the start and finish of the first stage of forging. The second stage forging simulations are depicted in Figures 35c and 35d. It may be observed that the first stage of forging is similar to pure compression, while the flow of material in the second stage of forging is more complex. As a result, the second stage of the IBR disk forging (Figure 36) is selected for demonstrating the methodology proposed in this work.

   A hypothetical rate dependent material with constitutive equation $\bar{\sigma} = \mathbf{k}\dot{\bar{\varepsilon}}^g \bar{\varepsilon}^h$ is used in the simulation.[43] In this case, the proportionality constant k, the strain rate exponent g, and the strain sensitivity factor h, are

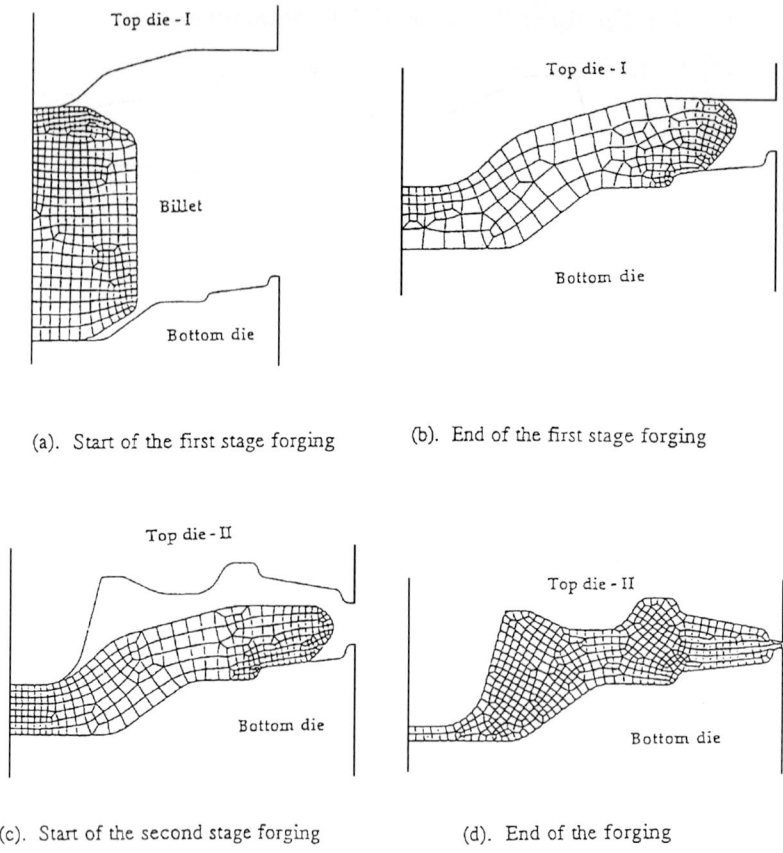

(a). Start of the first stage forging   (b). End of the first stage forging

(c). Start of the second stage forging   (d). End of the forging

Figure 35. Simulation of IBR disk forging.

assumed to be 10.0, 0.1, and 0, respectively. Due to the axisymmetric nature of the disk only one half of it is modeled and analyzed. The discretized workpiece has 302 quadrilateral elements and 347 nodes. The top die and bottom die are modeled using 203 elements (240 nodes) and 150 elements (176 nodes), respectively. An initial billet temperature of $1700°F$ and initial die temperature of $700°F$ (both top and bottom dies) are used for the simulation. The bottom die is kept stationary, while the top is die moved downwards to deform the material. Interface frictional conditions are enforced using a constant shear friction factor of 0.15.

The design requirement is to maintain the effective strain rate of element 275 at 0.5/s. In addition the temperature range in the workpiece is to be

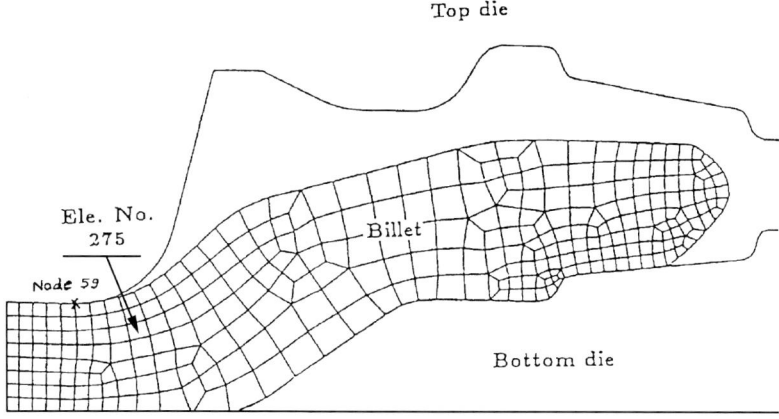

Figure 36.   Finite element model of IBR disk forging.

minimized. Element 275 is chosen as the 'control' element because it is observed that the strain rate of this element is largely varying in nature under normal simulation conditions (Figure 36). Therefore, it is a 'good' element to demonstrate the strain rate control strategy developed in this work. The design process is initiated with a die velocity of 1.0 in/sec. The simulation is carried out for 200 time steps and not until completion, because the goal of this example is limited to demonstrating the design methodology developed.

The optimization problem may be posed as,

$$\text{Minimize:} \quad (T_{\max} - T_{\min})$$

$$\text{Subject to:}$$

$$0.5 \le \dot{\varepsilon}_{275}$$

$$\dot{\varepsilon}_{275} \le 0.5$$

$$\text{Design variable:} \quad V_d$$

By repetitive simulations it is observed that node 59 has the lowest temperature in the workpiece for the entire duration of the simulation. The temperature range is therefore approximated as $(T_{\max} - T_{59})$, where, T_{59} is the temperature of node 59. Solving the above optimization problem results in optimal die velocity and strain rate trajectories for the process. The results

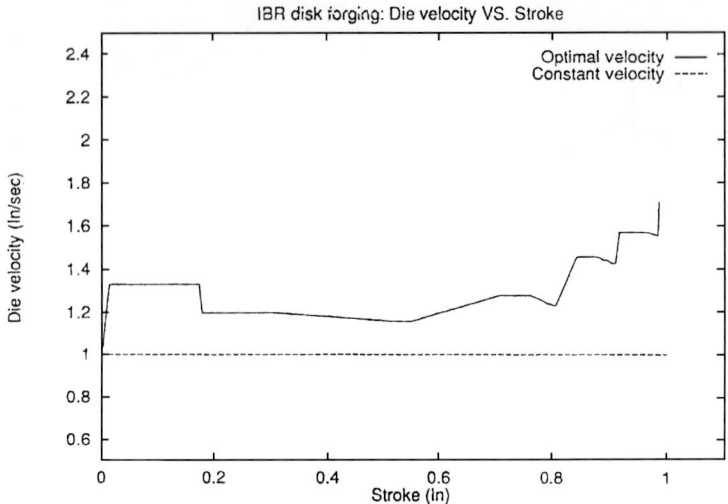

Figure 37. IBR disk forging: Die velocity profiles.

are compared with a constant velocity simulation, with a die velocity of 1.0 in/sec, to demonstrate the effectiveness of the optimization based design scheme.

The designed (optimal) ram velocity schedule and the corresponding strain rate trajectory (for element 275) are shown in Figures 37 and 38

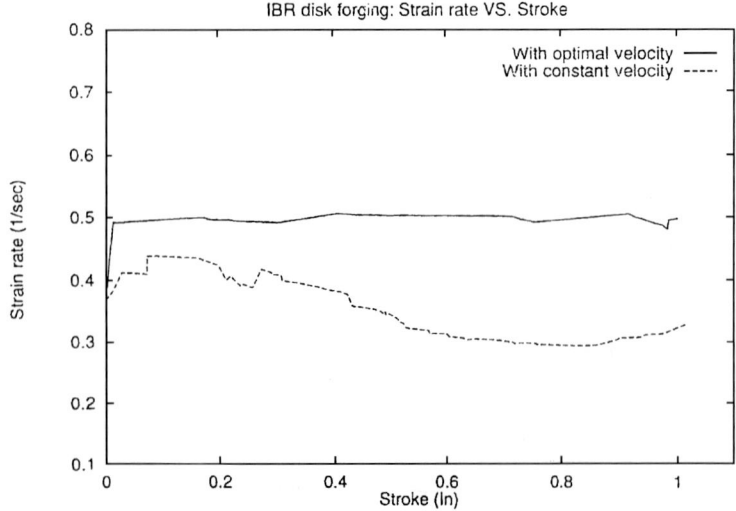

Figure 38. IBR disk forging: Strain rate trajectories.

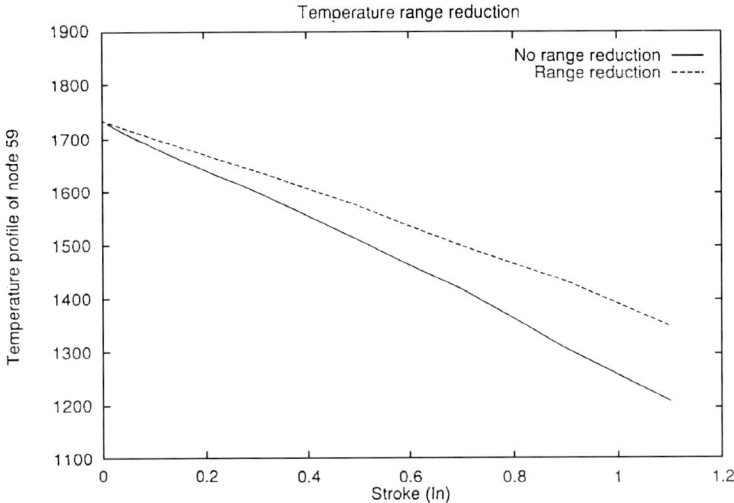

Figure 39. IBR disk forging: Nodal temperature profiles.

respectively. Figure 39 depicts the temperature path of node 59 for the optimal and constant ram velocity cases. It may be observed that under the optimal velocity condition the strain rate of element 275 is maintained for the entire duration of the simulation. Also, the temperature gradient at node 59 is considerably reduced (Figure 39). At the end of the simulation T_{59} is increased from $1230°F$ (for the constant velocity case) to $1355°F$ (for the optimal velocity case). The corresponding temperature range in the workpiece, at the end of the simulation, is reduced from $535°F$ to $392°F$, a decrease of about 26%. Therefore, using the optimization-based methodology the temperature range in the workpiece is reduced, and the strain rate requirement on element 275 is also satisfied.

This example demonstrates the effectiveness of the optimization-based methodology in meeting the design requirement under different deformation and temperature constraints and confirms the integrity and stability of the proposed approach for large scale finite element problems.

6.7. SUMMARY AND CONCLUSIONS

The forging of complex geometries using expensive alloys under optimum processing conditions is critical in manufacturing defect-free, cost-effective products on a repeatable basis. This work presents an innovative methodology for process modeling and 'control' using a multidisciplinary approach based on nonlinear finite element methods and mathematical programming techniques.

Earlier the authors used an approach based on optimal control theory for process parameter design. This method involved the building of a state space representation of the metal forming finite element equations linearized at a given velocity. The die velocity schedule and initial billet temperature were designed by minimizing a quadratic performance index at each time step. This involved generation of plant and input matrices, which needed extensive matrix operations and computations. To avoid these computationally demanding steps, in this work, the nonlinear finite element equilibrium equations are solved for element strain rates and node temperatures using a methodology based on numerical optimization. A standard mathematical optimization problem is formulated by placing constraints on strain rates and temperatures, with the die velocity schedule and initial billet temperature as design variables. More uniform deformation is obtained by minimizing the strain rate variance and/or temperature range in the workpiece. Approximation methods are used in solving the optimization problem to make the procedure more computationally efficient. Finite difference sensitivity analysis is performed, and the feasible directions algorithm is employed in designing optimal die velocity schedules and initial die temperature for the process. The formulation though general for plane strain and axisymmetric situations does require some problem specific changes.

A general purpose computer program has been developed to implement the methodology developed. Several design problems have been solved to substantiate and validate the design approach. Both strain rate and nodal temperature have been effectively controlled to satisfy the design requirements and improve the performance of the process.

From this study, the following conclusions can be drawn:

1. Optimal die velocity profiles may be used to effectively control the strain rate in the workpiece, as well as reduce the temperature gradient and temperature range for any forging process.
2. Optimizing the initial die temperature directly influences the die-workpiece boundary temperature and may be used to reduce the temperature range in the billet. Appropriate selection of initial die temperature can also considerably alleviate the die-chilling effect at the die-billet interface.
3. In some cases, additional benefits such as a reduction in die loads and process time were also observed.
4. Forging is a complex process with many interacting variables. For very complex geometries the design of die velocity and initial die/billet temperatures may not be sufficient to effectively control the process variables. Shape of the die (inclusive of the preform die shape) is a

major factor in the deformation behavior of the workpiece. Various defects such as localized deformation and improper die fill can be corrected by optimizing the preform die shape. Most of the earlier research in preform design was based on the backward tracing process where the simulation is started from the die fill position and the path taken in filling the die cavity is traced backwards. But this approach is purely theoretical in nature and physically infeasible. Future research in this field needs to address preform die design using standard forward simulations, probably through progressive and iterative modification of the die shape. Once this is accomplished, preform design and process parameter design could be integrated and posed as a simultaneous design optimization problem.

References

1. Lahoti, G.D. and T. Altan, 1979, "Research to develop process models for producing a dual property Titanium alloy compressor disk." *AFML-TR-79-4156*, Wright Patterson Air Force Base, Dayton, Ohio.
2. Devadas, C., I.V. Samarasekera, and E.B. Hawbolt, Feb 1991, "The thermal and metallurgical state of steel strip during hot rolling: Part III, Microstructural Evolution." *Metallurgical Transactions*, **22A**, 335–348.
3. Sastry, S.M.L., R.J. Lederich, T.L. Mackay and W.R. Kerr, 1983, "Generalized relations between metallurgical and process parameters for superplastic forming of Titanium alloy." *Experimental Verification of Process Models*, pp. 70–93. American Society for Metals.
4. Dadras, P. and J.F. JR. Thomas, Nov 1981, "Characterization and modeling for forging deformation of Ti-6Al-2Sn-4Zr-2Mo-0.1Si," *Metallurgical Transactions*, **12**A, 1867–1875.
5. Cohen, R.E. and D.R. Durham, 1990, "Microstructure as a criterion for the selection of hot working process parameters for plain medium carbon steel," *Journal of Materials and Technology*, **112**, 90–94.
6. Seetharaman, V., J.C. Malas, and C.M. Lombard, 1991, "Hot extrusion of a Ti-Al-Nb-Mn alloy," *Materials Research Society Proceedings*, **213**, 889–891.
7. Blazynski, T.Z., 1976, *Metal Forming*, pp. 20. John Wiley & Sons.
8. Altan, T., S.I. Oh, and H.L. Gegel, "Metal Forming," ASM, 136–141.
9. Avitzur, B., 1980, *Metal Forming, The Application of Limit Analysis*, pp. 137, 152 and 160. Marcel Dekker, Inc., New York.
10. Shabaik, A.H., 1978, "Analysis of Forming Processes: Experimental and Numerical Methods." *Applications of Numerical Methods to Forming Processes*, AMD-Vol. 28, ASME Winter Meeting, December, pp. 15–25.
11. Shahaf, M., M. Bercovier, D. Guez, and I. Blades, 1982, "Interactive simulation of a forging process for blades." *Numerical Methods in Industrial Forming Process*, pp. 343–350.
12. Kudo, H., 1960, "Some analytical and experimental studies of axisymmetric cold forging and extrusion–I." *Int. J. Mech. Sci.*, **2**, 102.
13. Kudo, H., 1960, "Some analytical and experimental studies of axisymmetric cold forging and extrusion–II." *Int. J. Mech. Sci.*, **3**, 91.
14. Nagpal, V., Nov 1974, "General kinematically admissible velocity fields for some axisymmetric and metal forming problems." *ASME Trans., Journal of Engineering for Industry*, 1197–1201.
15. McDermott, R.P. and A.N. Bramley, 1974, "An elemental upper-bound technique for general use in forging analysis." *Proc. 15th M.T.D.R. Conference*, 437.

16. Oh, S.I. and S. Kobayashi, 1975, "An approximate method for a three dimensional analysis of rolling." *Int. J. Mech. Sci.*, **17**, 293.

17. Zienkiewicz, O.C. and P.N. Godbole, 1974, "Flow of plastic and viscoplastic solids, with special reference to extrusion and forming process." *Int. J. Num. Meth. Engg.*, **8**, 3–16.

18. Zienkiewicz, O.C., P.C. Jain, and E. Onak, 1978, "Flow of solids during forming and extrusion: Some aspects of numerical solutions." *Int. J. Solids and Structures*, **14**, 15–38.

19. Lee, C.H. and S. Kobayashi, 1973, "New solutions to rigid plastic deformation problems using a matrix method." *Trans. ASME, J. Engr. for Ind.*, **95**, 865–873.

20. Kobayashi, S., 1977, "Rigid plastic finite element analysis of axisymmetric metal forming process." *Numerical Modeling of Manufacturing Process*, ASME PVP-PB-025, pp. 49–68.

21. Oh, S.I., N. Rebelo, and S. Kobayashi, 1978, "Finite element formulation for the analysis of plastic deformation of rate-sensitive materials in metal forming." *Metal Forming Plasticity — IUTAM Symposium*, pp. 273–291. Tutzing, Germany.

22. Oh, S.I., G.D. Lahoti, and T. Altan, 1982, "Application of finite element method to industrial metal forming process." *Numerical Methods in Industrial Forming Process*, pp. 145–153. Pineridge Press Ltd., Swansea, U.K.

23. Oh, S.I., G.D. Lahoti, and T. Altan, May 1981, "ALPID — A general purpose FEM program for metal forming." *Proceedings of NAMRC-IX, State College, Pennsylvania*, 83.

24. Kobayashi, S., S.I. Oh, and T. Altan, 1989, "Metal forming and the finite element method." Oxford University Press.

25. Dexter, R.J., K.S. Chan, and W.H. Couts, "Elastic-viscoplastic finite element analysis of a forging die." *Int. J. Mech. Sci.*, **33**(8), 659–674.

26. Haque, I., J.E. Jr. Jackson, T. Gangjee, and A. Raikar, 1987, "Empirical and finite element approaches to forging die design: A state of the art survey." *J. Materials Shaping Technology*, **5**(1), 23–33.

27. Wu, W.T. and S.I. Oh, 1985, "ALPIDT: A general purpose FEM code for simulation of nonisothermal forming processes." *Proc. NAMRC-XIII Conference*. Berkeley, California.

28. Oh, S.I., J.J. Park, S. Kobayashi, and T. Altan, 1983, "Application of FEM modeling to simulate metal flow in forging a Titanium alloy engine disk." *Trans. ASME, J. Engr. Ind.*, **105**, 251.

29. Zienkiewicz, O.C., E. Onate, and J.C. Heinrich, 1978, "Plastic flow in forming, I. Coupled thermal behavior in extrusion. II. Thin sheet forming." *Applications of Numerical Methods to Forming Processes*, ASME, AMD, **28**, 107.

30. Tang, M.C. and S. Kobayashi, 1982, "An investigation of the shell nosing process by the finite element method, Part 2: Nosing at elevated temperatures (Hot nosing)." *J. of Engr. for Industry, Trans. ASME*, **104**(3), 312–318.

31. Argyris, J.H. and J. Doltsins, Sr., 1981, "On the natural formulation and analysis of large deformation coupled thermal mechanical problems." *Comp. Meth. Appl. Mech. Eng.*, **25**, 195.

32. Antares User's Manual, UES, Inc.. Dayton, Ohio.

33. Boer, C.R., H. Rydstad, and G. Schroder, 1985, "Choosing optimal forging conditions in isothermal and hot-die forging." *J. Applied Metalworking*, **3**(4), 421–431.

34. Lanka, S., R. Srinivasan, and R.V. Grandhi, Sept 1991, "A design approach for intermediate die shapes in plane strain forging." *J. Mater. Shaping Technol.*, pp. 193–206.

35. Malas, J.C., 1985, "A thermodynamic and continuum approach to the design and control of precision forging processes." M.S. Thesis, Wright State University, Dayton, Ohio.

36. Hong, B.S., 1990, "Numerical simulation of stability control in sheet metal forming," M.S. Thesis, Massachusetts Institute of Technology, Cambridge, Massachusetts.

37. Park, J.J., N. Rebelo, and S. Kobayashi, 1983, "A new approach to preform design in metal forming with the finite element method." *Int. J. Machine Tool Des. Res.*, **23**, 71.

38. Han, S.C., R.V. Grandhi, and R. Srinivasan, 1993, "Optimum design of forging die shapes using nonlinear finite element analysis." *AIAA Journal*, **31**, 774–781.

39. Srinivasan, R., G.H.K. Reddy, S.S. Kumar, and R.V. Grandhi, August 1994, "Intermediate shapes in closed-die forging by the backward deformation optimization method (BDOM)." *Journal of Materials Engineering and Performance*, **3**(4), 501–513.

40. Grandhi, R.V., A. Kumar, A. Chaudhary, and J.C. Malas, 1993, "State space representation and optimal control of nonlinear material deformation using the finite element method." *Int. J. Num. Meth. Engg.*, **36**, 1967–1986.

41. Kumar, A., R.V. Grandhi, A. Chaudhary, and D. Irwin, 1993, "Modeling and design of control parameters in metal forming processes". *Transactions, Canadian Society of Mechanical Engineers*, **17**(4A), 613–631.

42. Cheng, H., R.V. Grandhi, and J.C. Malas, 1994, "Design of optimal process parameters for non-isothermal forging". *International Journal for Numerical Methods in Engineering*, **37**, 155–177.

43. "MME-ALPID User's Manual," UES, Inc., Processing Science Division. Dayton, Ohio.

44. Grandhi, R.V., R. Thiagarajan, J.C. Malas, and D. Irwin, 1995, "Reduced-order state space models for control of metal forming processes." *Optimal Control Applications & Methods*, **16**, 19–39.

45. DOT User's Manual, 1993, *VMA Engineering*. Goleta, California.

46. DOC User's Manual, 1993, *VMA Engineering*. Goleta, California.

47. Vanderplaats, G.N., 1984, *Numerical Optimization Techniques for Engineering Design: With Applications*. McGraw-Hill, Inc.

48. Haftka, R.T. and Z. Gurdal, 1992, *Elements of Structural Optimization*. Kluwer Academic Publishers, Netherlands.

49. Chapra, S.C. and R.P. Canale, 1988, *Numerical Methods for Engineers*. McGraw-Hill Book Company, U.S.A.

50. Arora, J.S., 1989, *Introduction to Optimum Design*. McGraw-Hill Book Inc., USA.

51. Schmit, L.A. Jr., and B. Farshi, 1974, "Some approximation concepts for structural synthesis." *AIAA Journal*, **12**(5), 692–699.

17696